Safety and Health Division 2014

Core Programming Area at the 2014 AIChE Spring Meeting and 10th Global Congress on Process Safety

New Orleans, Louisiana, USA
30 March - 3 April 2014

ISBN: 978-1-63439-078-1

Printed from e-media with permission by:

Curran Associates, Inc.
57 Morehouse Lane
Red Hook, NY 12571

Printed by Curran Associates, Inc. (2014)

For permission requests, please contact AIChE
at the address below.

AIChE
3 Park Avenue
New York, NY 10016-5991

Phone: (203) 702-7660
Fax: (203) 775-5177

www.aiche.org

Additional copies of this publication are available from:

Curran Associates, Inc.
57 Morehouse Lane
Red Hook, NY 12571 USA
Phone: 845-758-0400
Fax: 845-758-2634
Email: curran@proceedings.com
Web: www.proceedings.com

TABLE OF CONTENTS

The Growth of Idea Ownership: Patent Infringement and Risk Management within the Chemical Process Industries

Paul LaVanway, Esq.
Fredrikson & Byron, P.A.

14AIChE Spring Meeting, New Orleans, LA

Fredrikson & Byron, P.A.

Overview of Presentation

I. Patent Activity Update

II. Practical Risk Management Techniques

III. New Tools

14AIChE Spring Meeting, New Orleans, LA

Fredrikson & Byron, P.A.

PATENT ACTIVITY UPDATE

14 AIChE Spring Meeting, New Orleans, LA

Fredrikson & BYRON, P.A.

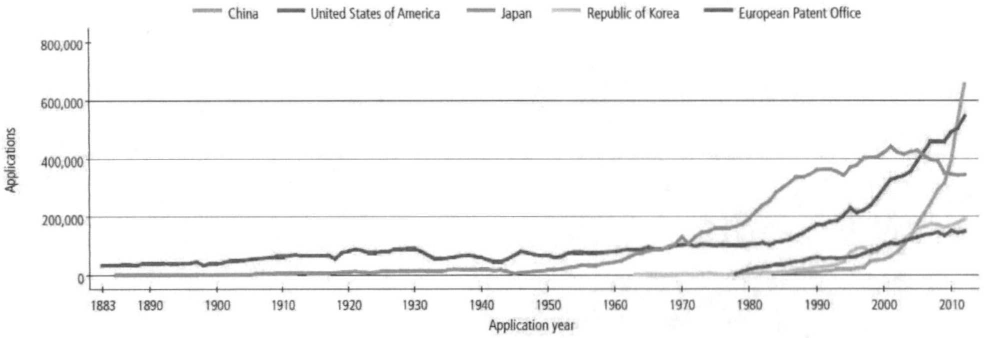

14 AIChE Spring Meeting, New Orleans, LA

Fredrikson & BYRON, P.A.

Worldwide Filing Trends

Origin	Field of technology															
	Australia	Canada	China	Finland	France	Germany	Italy	Japan	Netherlands	Republic of Korea	Russian Federation	Sweden	Switzerland	United Kingdom	United States of America	Others
Chemistry																
Organic fine chemistry	1,164	2,798	18,900	539	20,317	38,897	4,436	38,136	6,148	10,424	2,185	5,375	17,365	12,314	75,204	30,237
Biotechnology	2,592	3,432	17,402	804	8,107	16,915	2,306	20,814	5,500	9,165	1,822	1,765	7,316	7,058	67,521	21,840
Pharmaceuticals	3,647	6,297	38,194	958	18,328	31,204	7,370	28,798	6,362	10,514	5,692	7,748	23,615	15,563	123,173	50,208
Macromolecular chemistry, polymers	382	816	11,135	2,173	4,439	20,009	2,446	45,323	4,576	7,906	935	325	3,439	1,714	31,810	7,538
Food chemistry	875	1,122	19,191	428	2,595	5,330	1,070	13,272	6,409	12,111	17,015	337	4,744	2,464	19,963	11,149
Basic materials chemistry	1,002	1,992	24,266	901	6,012	35,527	1,536	42,724	7,038	11,113	3,343	690	6,970	6,340	51,100	13,535
Materials, metallurgy	1,697	1,569	30,285	1,729	6,697	17,150	1,480	43,685	1,846	12,235	8,359	1,514	2,186	2,422	20,753	12,182
Surface technology, coating	670	1,386	11,959	1,185	4,907	15,954	1,642	52,591	1,933	10,246	1,779	1,200	2,604	2,498	36,858	7,336
Micro-structural and nano-technology	116	96	1,710	124	606	1,371	94	2,405	254	2,579	587	134	153	158	2,486	683
Chemical engineering	1,479	2,235	16,952	1,840	7,352	25,809	2,846	32,504	4,788	13,354	4,426	2,422	4,287	5,490	40,121	13,512
Environmental technology	762	1,514	13,620	855	5,042	13,594	1,481	27,259	2,533	12,888	2,303	1,116	1,765	2,719	20,263	8,002

14AIChE®
Spring Meeting, New Orleans, LA

Fredrikson & BYRON, P.A.

Worldwide Filing Trends

Field of technology	Publication year					Growth rate
	2007	2008	2009	2010	2011	2007-11 (%)
Chemistry						
Organic fine chemistry	52,609	53,873	53,009	52,559	51,271	-0.6
Biotechnology	33,929	35,654	37,652	38,423	40,849	4.7
Pharmaceuticals	73,345	73,769	72,126	69,258	69,311	-1.4
Macromolecular chemistry, polymers	27,627	28,345	28,800	28,683	28,748	1.0
Food chemistry	21,690	23,582	27,245	27,889	30,858	9.2
Basic materials chemistry	38,920	41,085	42,322	43,795	45,115	3.8
Materials, metallurgy	30,200	33,775	34,968	36,988	38,542	6.3
Surface technology, coating	29,348	30,732	32,792	33,215	33,954	3.7
Micro-structural and nano-technology	2,302	2,478	2,829	3,094	3,203	8.6
Chemical engineering	33,263	35,087	35,766	36,618	37,991	3.4
Environmental technology	21,688	22,605	24,333	25,572	26,324	5.0

14AIChE®
Spring Meeting, New Orleans, LA

Fredrikson & BYRON, P.A.

Comparative Trends

Field of technology	Publication year					Growth rate
	2007	2008	2009	2010	2011	2007-11 (%)
Electrical engineering						
Electrical machinery, apparatus, energy	101,013	105,591	111,876	116,009	122,697	5.0
Audio-visual technology	92,978	91,258	85,411	80,607	75,881	-5.0
Telecommunications	68,915	69,804	60,808	56,417	49,533	-7.9
Digital communication	57,696	64,237	69,191	74,389	79,726	8.4
Basic communication processes	17,208	17,734	17,164	16,515	15,554	-2.5
Computer technology	123,504	134,493	133,072	130,052	134,272	2.1
IT methods for management	19,370	21,684	25,022	23,261	23,532	5.0
Semiconductors	76,617	81,238	78,685	77,547	80,049	1.1
Mechanical engineering						
Handling	42,764	42,783	42,764	42,414	44,344	0.9
Machine tools	36,607	38,315	40,585	43,242	46,363	6.1
Engines, pumps, turbines	41,891	43,573	48,156	48,372	48,423	3.7
Textile and paper machines	35,815	33,681	32,251	30,633	30,364	-4.0
Other special machines	44,008	45,912	47,541	49,011	50,981	3.7
Thermal processes and apparatus	25,280	25,699	27,259	29,382	29,855	4.2
Mechanical elements	43,957	47,513	47,286	46,405	46,913	1.6
Transport	65,237	67,717	70,542	67,015	65,439	0.1

14AIChE®
Spring Meeting, New Orleans, LA

Fredrikson & BYRON, P.A.

Change in 5 year average over last 20 Years

Fastest Growing

Category	Growth Rate
Computer Software	+11%
Data Processing	+11%
Semiconductor Devices	+10%
Video Distribution	+10%
Computer Hardware	+8%
Chemical – Crystals	+8%
Nanotechnology	+8%
Information Storage	+6%
Communications	+6%
Design	+5%

Slowest Growing

Category	Growth Rate
Chemical – Purification/Evaporation /Distillation	-2%
Chemical – General Compounds	-3%
Time Measurement	-3%
Machine Mechanism	-3%
Chemical Manufacture Specific	-3%
Organic Compounds	-3%
Pipes & Joints	-4%
Education	-4%
Hazardous Waste	-4%
HVAC	-4%

14AIChE®
Spring Meeting, New Orleans, LA

& BYRON, P.A.

US Patent Filing Trends

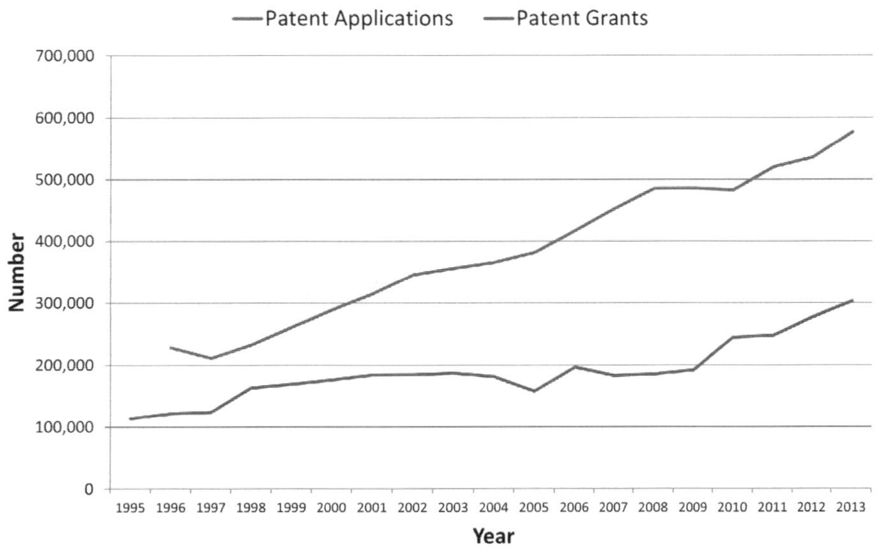

14 AIChE® Spring Meeting, New Orleans, LA

Fredrikson & BYRON, P.A.

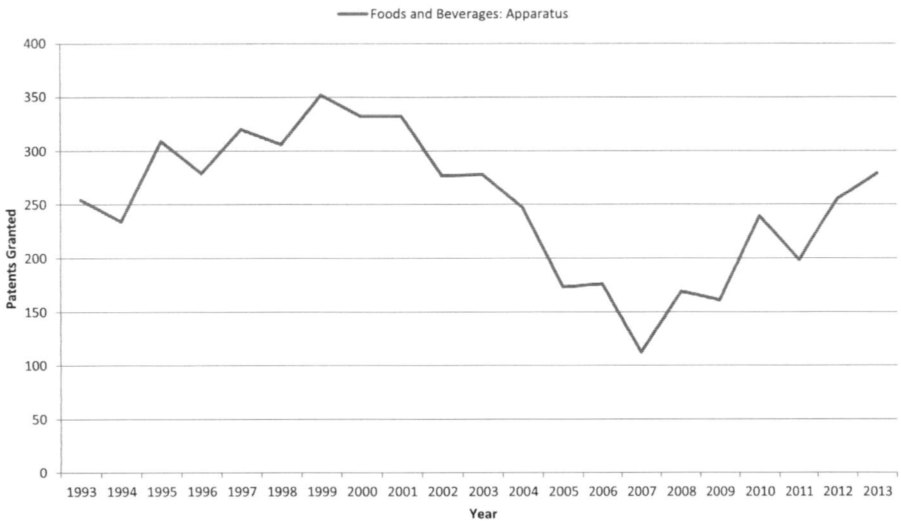

14 AIChE® Spring Meeting, New Orleans, LA

Fredrikson & BYRON, P.A.

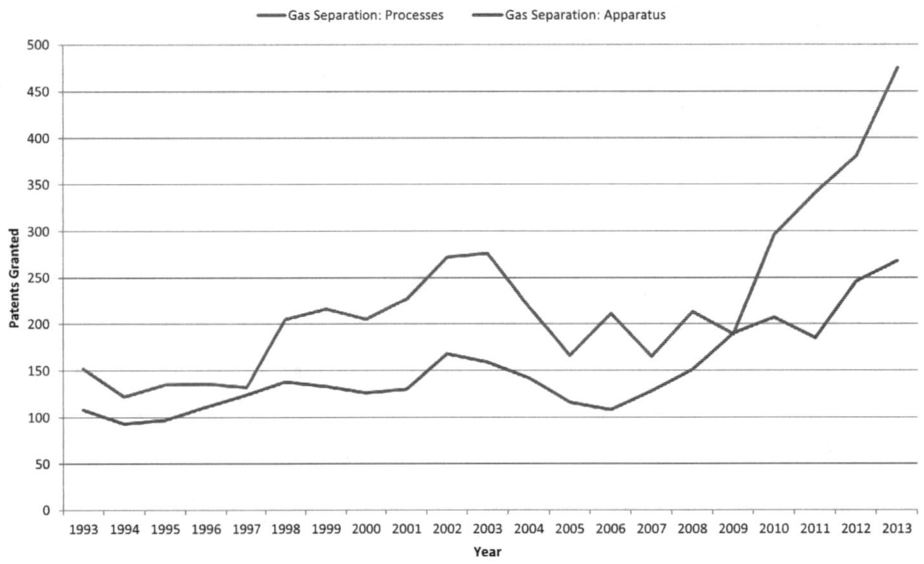

14AIChE®
Spring Meeting, New Orleans, LA

Fredrikson & BYRON, P.A.

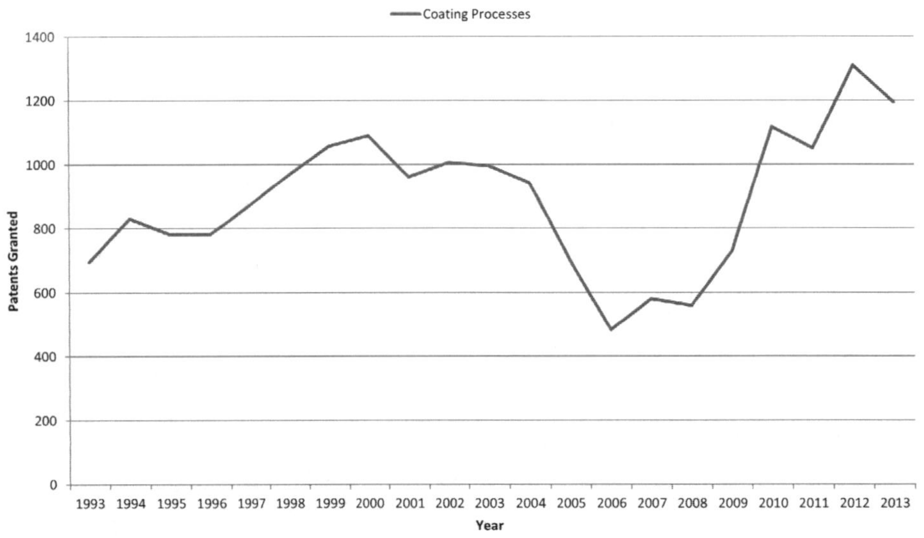

14AIChE®
Spring Meeting, New Orleans, LA

Fredrikson & BYRON, P.A.

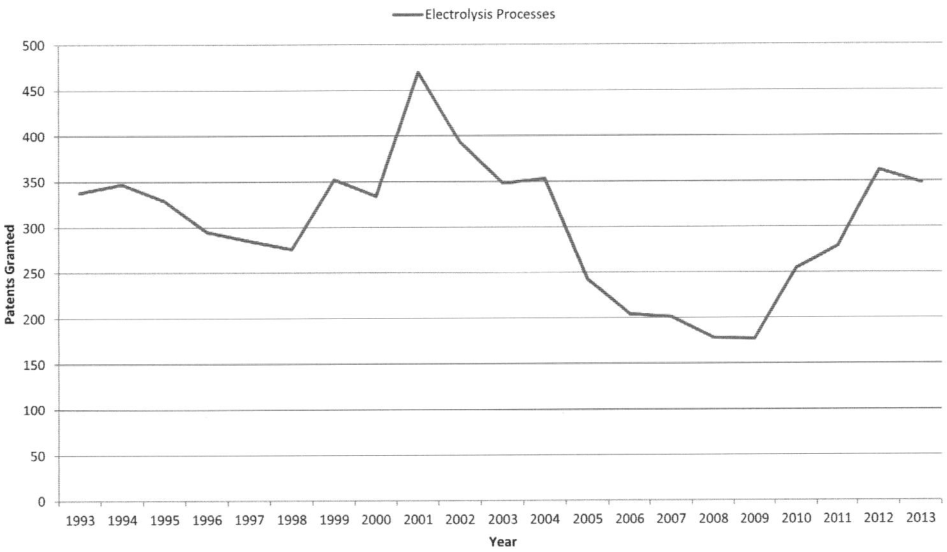

14 AIChE®
Spring Meeting, New Orleans, LA

Fredrikson & BYRON, P.A.

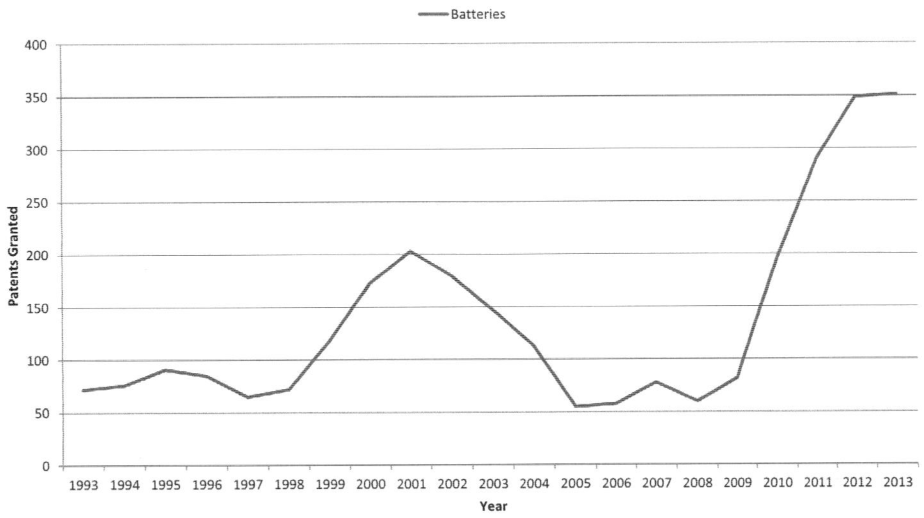

14 AIChE®
Spring Meeting, New Orleans, LA

Fredrikson & BYRON, P.A.

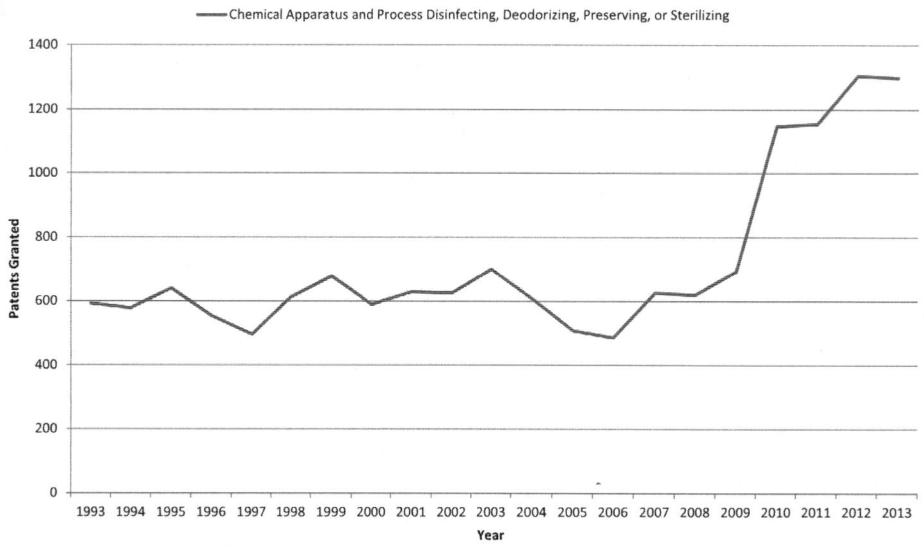

14 AIChE®
Spring Meeting, New Orleans, LA

Fredrikson & BYRON, P.A.

Patent Litigation Statistics

- Median damage award last 15 years all cases
  - $1.9 M → 16.5 M
- Median over last 5 years
  - $4.9M
- Chemical Industry (excluding bio/pharma)
  - Generally smaller median damages (less than $3M) but characterized by outliers
- Disruptive
  - Median time to trial 2.3 years

14 AIChE®
Spring Meeting, New Orleans, LA

Fredrikson & BYRON, P.A.

Cost of Patent Infringement Litigation

Amount at Risk	Median Cost
Less than $1 Million	$700,000
$1-$10 Million	$2,000,000
$10-$25 Million	$3,325,000
$25 Million+	$5,500,000

14AIChE® Spring Meeting, New Orleans, LA

Fredrikson & BYRON, P.A.

Distribution of Patent Cases

Top Ten Industry	Percent of Cases
Consumer Products	18%
Biotechnology / Pharma	13%
Industrial / Construction	11%
Computer Hardware / Electronics	9%
Medical Devices	9%
Business / Consumer Services	6%
Software	6%
Automotive / Transportation	5%
Telecommunications	5%
Chemical / Synthetic Materials	5%

14AIChE® Spring Meeting, New Orleans, LA

Fredrikson & BYRON, P.A.

But . . . Looking at 11-15

Rank	Industry
11	Food/Beverages/Tobacco
12	Metals/Mining
13	Clothing/Textiles
14	Energy
15	Agriculture

14 AIChE®
Spring Meeting, New Orleans, LA

Fredrikson & BYRON, P.A.

RISK MANAGEMENT STRATEGIES

14 AIChE®
Spring Meeting, New Orleans, LA

Fredrikson & BYRON, P.A.

Understanding the Nature of the Risk

- "Fuzzy" boundaries Problems

<u>Features</u>

- ~~Back~~
- Seat
- ~~Four~~ Legs resting on the ground

<u>Revised</u>

- Seat
- At least one base contacting the ground

14 AIChE Spring Meeting, New Orleans, LA

Fredrikson & BYRON, P.A.

Understanding the Nature of the Risk

<u>Revised</u>

- Seat
- At least one base contacting the ground

Legal Scope Not Necessarily Co-extensive with Thrust of Description

14 AIChE Spring Meeting, New Orleans, LA

Fredrikson & BYRON, P.A.

Managing New Product Risk

- Correlate Effort to Risk
  - Repeat Use of Old Technology in Existing Field
  - Old Technology in Different Field
  - New Technology in Existing Field
  - New Technology in New Field

Increasing Risk

14 AIChE Spring Meeting, New Orleans, LA

Fredrikson & BYRON, P.A.

Other Factors to Consider

- Intensity of Competition within Field
- Past History Between Competitors
- Value of Marketplace and Opportunity
- Relative Size of Parties
- Monetization Proposition

14 AIChE Spring Meeting, New Orleans, LA

Fredrikson & BYRON, P.A.

Managing New Product Risk

- Use a Phased Approach
  - Phase I
    - Basic Understanding of Landscape and Any "Show Stopping" Patents
    - Focus on Likely Competitors – Review Patents, Literature, Products
  - Phase II
    - Revisit Base Design and Evaluate Ancillary Features
  - Phase III
    - Check Any Design Modifications or Newly Issuing Patents

14 AIChE Spring Meeting, New Orleans, LA

Fredrikson & BYRON, P.A.

Managing Ongoing Risk

- Monitor Key Competitors / Inventors
- Monitor Technology Space
- Work with Sales and Marketing Personnel to Track New Product Introduction
- Watch Partners and Consultants

14 AIChE Spring Meeting, New Orleans, LA

Fredrikson & BYRON, P.A.

NEW TECHNIQUES FOR ACTION

14 AIChE® Spring Meeting, New Orleans, LA

Fredrikson & BYRON, P.A.

Knowing What Tools Are Available

- Pre-Issuance Submission
  - Attack the Patent Application Before it Becomes a Problem
- Simple But Limited Patent Office Attack
  - Ex Parte Reexamination
- More Involved Patent Office Attacks
  - Inter Partes Review
  - Post Grant Review

14 AIChE® Spring Meeting, New Orleans, LA

Fredrikson & BYRON, P.A.

CONCLUSIONS

- CPI Patents Generally Trending Upward
- Great Sector Specific Variability
- Effective Risk Management Requires Being Proactive and Deploying Resources Efficiency
- New Tools to Attack Problem Patents

14AIChE
Spring Meeting, New Orleans, LA

Fredrikson & BYRON, P.A.

QUESTIONS?

14AIChE
Spring Meeting, New Orleans, LA

Fredrikson & BYRON, P.A.

Chemical Engineering & the Law --Case Examples in Industrial Litigation

Paper# 352190

Satish Almaula
Senior Principal Consultant
Baker Engineering and Risk Consultants, Inc.
salmaula@bakerrisk.com

Presented at American Institute of Chemical Engineers Spring Meeting, New Orleans LA April 2014

Abstract

Chemical, petroleum, and allied manufacturing sectors involved in research, technologies, engineering and project management at times encounter issues and problems in successfully meeting a project's intended financial, technical, and operational goals and expectations. Significant deviations from goals and expectations can result in lawsuits between involved parties. Alternatively, parties file insurance claims, which can also result in lawsuits.

Research and technology issues can result in patent and intellectual property legal cases. In addition, issues can arise from engineering design, licensing, product liability, safety, regulatory compliance, construction and operational deficiencies, or an inability to meet contractual and performance commitments. Accidents and incidents at manufacturing or other facilities can result in damage to property and the environment, injuries and death as well as interruption to normal operation causing business loss. Such damages and losses also can result in lawsuits and insurance claims.

The need for collaboration between lawyers, scientists, engineers and other technical specialists is obvious when conducting legal, technical and business analyses to get to the root cause of the matter to support or defend a legal complaint or an insurance claim.

This paper presents a brief introduction to lawsuits and how legal and technical teams can collaborate with their respective experience and expertise in handling a lawsuit either as a plaintiff's or a defendant's team.

The main focus of the presentation will be case examples that demonstrate how principles familiar to engineers and other technical specialists play a part in litigation involving diverse issues in the chemical, refining and allied business sectors.

1 Introduction

Technical professionals, whether they are engineers, scientists or technical specialists often get drawn into legal disputes. Modern day media reports of claims and settlements tend to focus heavily on events in the electronic, information technology, pharmaceutical and drug industries, especially when the issues have a direct effect on consumers and involve huge sums of money.

Petroleum, chemical processing, and allied manufacturing sectors have their share of claims and disputes involving large amounts of money but do not always get the high visibility as the above-mentioned industries. Catastrophic incidents such as fires, explosions, release of hazardous materials leading to deaths, injuries and property damage do get expanded coverage and high visibility.

It is not surprising, therefore that chemical engineers, scientists and technology specialists are often drawn into insurance claims and legal disputes. Lawyers, insurers, scientists, engineers, and technical specialists often collaborate in such matters.

This presentation provides a brief background on typical issues leading to legal disputes followed by selected case examples. Names of organizations, involved parties, and individuals are not disclosed or discussed in this presentation. The selected examples are from the author's experience in lawsuits related to chemical processing and allied industries.

2 Insurance Claims and Lawsuits

Insurance claims and lawsuits can arise from diverse losses and disputes such as property damage, fatalities and injuries, negligent performance or, in extreme but uncommon instances, they can arise from intentional or criminal acts.

Losses alleged in processing industry claims and disputes tend to be related to: patent and intellectual property; engineering design; licensing; product liability; safety; regulatory compliance; construction and operational deficiency; and the inability to meet contractual and performance commitments.

Technical professionals including those in chemical engineering and similar fields can be drawn into insurance claims and lawsuits as fact or expert witnesses.

Fact witnesses are generally limited to testifying about facts that they know firsthand about an issue without expressing an opinion about them. Expert witnesses are generally technical professionals with extensive experience and in-depth knowledge in a particular subject matter. They analyze facts, evidence and information and give testimony on their findings as well as their expert opinions.

Given the diversity of training and experience of chemical engineers, they can be called upon to serve as experts in correspondingly diverse fields, some of which may or may not fall under a classic view of chemical engineering.

Collaboration between insurers, lawyers and technical experts typically involves, but is not limited to, the following:

- Developing a clear understanding of the claims and allegations.
- Knowing and understanding the involved parties and the type of business they are in.
- Gathering the information, evidence and facts necessary for legal, technical and business analyses.
- Performing a screening analysis of the strengths and weaknesses of legal, technical and business factors to support or rebut the claims and allegations.
- Determining the strategies for the approach and steps to be implemented by the legal, technical and business professionals in handling the matter.

3 Selected Examples

Example 1 -- Patent Related Lawsuit

A producer of asphalt shingle roof products hired a retired former senior executive as a consultant to reduce disposal costs of reject materials from production. Roofing shingles typically contain asphalt, filler material and granules evenly placed on a core material such as fiber glass or cellulose. The produced material is cut into standard sizes called "a course." The cutting produces some unusable material ("rejects" or "rejected material").

The consultant researched, developed and pilot-tested a method to recycle rejects back into production. The reject material, containing asphalt and filler material, was finely ground with limestone and then recycled back into production, saving costs of filler material and asphalt and avoiding costs associated with the disposal of the reject material. The tests were successful and easily scaled up to commercial level.

The consultant asked the company to file a patent for the new recycling method, but the company procrastinated and, ultimately, did not file for a patent.

Later, the consultant, after ending his work with the company, filed and obtained a patent for the new recycling method. Upon learning of the patent, the company sued the consultant/inventor for stealing proprietary information, which initially prevented him from offering the technology to others. The viability, cost effectiveness, and worth of the new technology as well as information protection were all issues raised in the lawsuit.

Analysis showed that the proposed retrofit of an average size recycle plant at 70 plants in the industry across the U.S. would roughly total $156,000,000 in capital costs. Estimated annual savings to the industry from reduced raw material costs and avoidance of disposal costs amounted to $190,000,000, yielding an attractive return on investment. So, the viability and cost effectiveness of the new technology were demonstrated in principle.

The worth of the new technology/licensing value was determined by financial and economic experts.

The matter went to trial, and the consultant/inventor won the case.

Principles familiar to technical specialists applied in this example: Manufacturing processes, unit operations, industry survey, capital cost estimation, annual cost estimation, return on investment, and protection of important information.

Example 2 -- Technology Transfer, Design-Build, Project Management Lawsuit

The owner of an operating plant wanted to convert by-product waste to useful and saleable products. The owner acquired a license for new technology to be used for the conversion of waste to intermediates. The owner also acquired established technologies to convert intermediates to saleable products. The new technology licensor represented to the owner that the technology was demonstrated and ready for commercial use. The licensor also offered used equipment in good condition, which the owner agreed to purchase.

A design-build contract for the plant and process for the new technology was awarded on a fixed price ($150 million) and fixed schedule basis to an architect and engineering firm ("A&E").

The technology licensor provided a Process Information Package (PIP) for design-build work as "frozen" --- that is, ready for detailed design and procurement. Numerous and untimely changes were imposed by both the technology licensor and the owner, causing delays in the detailed design, procurement and project completion. The A&E submitted change orders to the owner, who orally approved them.

The technology sold to the owner was actually its first application on a commercial scale in the U.S. The only other reported use of the technology on a commercial scale was in Japan, where several modifications were needed to make the process work correctly. The condition of the used equipment sold to the owner was not "good," as claimed. The equipment required significant cleaning and modification to meet design and code requirements.

Project completion was delayed by three months but, when completed, the system operated satisfactorily. The A&E submitted a Request for Equitable Adjustment ("REA") to the fixed price for the additional costs. In response, the owner sued the A&E for "Bad Project Management" claiming damages amounting to seven times the REA amount. Of note is that the technology licensor was not drawn into the lawsuit and was not held accountable for changes made to the "frozen" PIP.

The matter went to mediation, but did not settle. The owner later filed for bankruptcy, and the plant was bought by another party.

Principles familiar to technical specialists applied in this example: Review of process design, PFDs, P&IDs, impact of design changes and delays on construction, project management, and contractual performance.

Example 3 -- Process Technology Licensing Lawsuit

Two separate companies (A and B) had plants located at one site. Company A manufactured product X1 and sent its wastewater to Company B for treatment and discharge. Company A then started to manufacture another product, X2, at the same plant. Company A recognized that Company B's plant could not handle the additional wastewater created by the process to manufacture X2. Therefore, Company A decided to build its own new treatment plant on the site to handle the wastewater generated by products X1 and X2 combined.

Company A chose a wastewater treatment contractor (TC) to provide "turn-key" services, including technology, engineering, design, construction and startup of the new wastewater treatment plant. The selected technology was a "thermophyllic" biological treatment process, patented and licensed by a third party technology developer ("Licensor"). The license to operate using the technology was granted to Company A, and the technology and engineering details were given by the Licensor to TC.

The Licensor pilot-tested Company A's waste stream to confirm treatability and provided TC with process and engineering information tailored to treat the tested waste stream samples.

At startup, the new wastewater treatment plant failed to meet the treatment targets, and Company A filed a lawsuit against TC. The Licensor was drawn into the lawsuit as a third party defendant.

During the discovery phase of the lawsuit, the allegations against the Licensor were determined to be unfounded both technically and contractually. It was established that there were numerous deficiencies in Company A's design basis as well as defects in both the engineering design and installation by TC. There also were issues involving the performance warranties between Company A and TC.

The technology provided by the Licensor had been commercially proven at other facilities, and it was established that the Licensor had met its contractual obligations.

The matter was settled before trial.

Principles familiar to technical specialists applied in this example: Process evaluation, pilot testing, process design and engineering analysis, contractual commitments, review and understanding of Company A's and TC's obligations and failures.

Example 4 -- Plant Incident and Alleged Medical Injury

A manufacturing plant was undergoing a maintenance turnaround with contractor employees working on scaffolding near the top of a large carbon steel tank containing concentrated sulfuric acid. An undetected water leak into the tank diluted the acid and a relief valve on top of tank momentarily lifted and released vapors with a sulfurous odor. The contractor employees immediately evacuated the area and reported the incident. The turnaround work was shut down for the day, and all workers were sent home.

Reportedly, the contractor employee who was closest to the top of the tank went to a restaurant with his co-workers, where he collapsed. He was attended to and recovered. Later, the contractor employee sued the plant for negligence, alleging that the gas he was exposed to was hydrogen sulfide (H_2S). During the litigation, medical statements were produced stating that the contractor employee's nervous system was permanently damaged from exposure to H_2S.

The defense argued that the released gas was more likely to be sulfur dioxide (SO_2) and not H_2S. Reported public information showed that exposure to SO_2 could have bronchial effects but not permanent nerve damage.

A literature search as well as an examination of the conditions of the carbon steel tank (pressure, temperature, and acid concentration) on the date and time of the incident suggested that the gas released was as a result of corrosion, and it likely contained primarily SO_2 with trace amounts of H_2S.

Defense lawyers engaged an independent third party laboratory to test evolution of gases from reaction of concentrated and diluted sulfuric acid on carbon steel, thus simulating conditions in the tank on the day and time of the release. The tests showed that the evolved gas content was mainly SO_2 with likely trace amounts of H_2S.

A chemistry professor for the plaintiff testified in his deposition that the gas released at the time of the incident was primarily H_2S, but his testimony was strongly challenged by defense lawyers. A second professor from another university testified in his deposition that the plant violated OSHA's Process Safety Management (PSM) regulation. The second professor's testimony was challenged on the basis of his lack of experience with PSM.

Later, it was learned that the contractor employee who filed the lawsuit had been previously employed as a sheet rock worker and had developed indications of dust-related issues involving his bronchial system.

The matter was settled between the parties before trial.

<u>Principles familiar to technical specialists applied in this example:</u> Chemistry of reaction of dilute acid with carbon steel; examination of the incident timeline and possible cause(s) of water backflow into the acid tank; basic metallurgy of carbon steel; review of third party test results; and review of the plant's PSM program.

4 Observations and Conclusion

1. Insurance claims and lawsuits can arise from a variety of reasons and situations, and outcomes are not easily predictable.
2. Lawsuits and insurance claims can arise from unwritten "perceived expectations" by the involved parties.
3. Besides the terms and conditions defined in a contract, other factors, such as standard of care and generally accepted good practice in a given industry, are often encountered in allegations.
4. In addition to directly involved parties, all other entities that are in any way associated with the matter may be drawn into the lawsuit - especially those perceived to have "deep pockets."
5. Claims of advantages and success using new technology on a commercial scale should be thoroughly evaluated before acquiring the license for its use.
6. Clear agreements between a company, its consultants or involved third parties regarding ownership and rights regarding intellectual property need to be executed in advance.
7. Applying fixed price and fixed schedule terms to design-build contracts in capital projects (or any) in which a firm and definitive scope cannot be described clearly in the contractual documents will most likely result in cost overruns and schedule delays, prompting lawsuits.
8. A technical professional can take steps to minimize liability, although it is doubtful one can completely avoid it. Some exemplar suggestions are:
    - Try to clearly understand the technical and (to the extent possible) commercial needs of an assignment.
    - Take on only what you are capable of performing on a sound technical basis. Do not make unsupportable assumptions or try to "wing it."
    - Speak up if you are not the best person to take on the assignment. Have it passed on to someone else who can best perform the work.
    - Keep good written factual records of your work and your communications about the work you are doing.
    - Most of all, be factual and truthful in all you do.

Determining The Liability Of A Fire Starting Product

C. J. Abraham
Scientific Advisory Services Ltd
Great Neck, New York

Abstract

Natural Fire Starter is an ethanol based fuel gel which was designed specifically for wood and wood pellet stoves. It is sold in a plastic bottle with a child proof closure. This product had never been cited in any reported accident until a 7-year old was badly burned when he was allowed, unsupervised, to toast marshmallows in a wood burning fireplace. The subsequent events reported by the child, his parents and brother are discussed as well as the experts and lawyers claiming the liability of the defendants. Said defendants include the manufacturer of the formulation, the bottle manufacturer, manufacturer of the closure and the packager.

In order to reconstruct the accident and determine the cause and liability involved, this investigator had to consider the testimony of the injured child, his parents, brother , the ENT personnel, medical personnel, plus testing and testimony of all of the experts as well as the statements made by the plaintiff's attorney.

Determining The Liability Of A Fire Starting Product

As simple as the product appears, over 50 characteristics and variables of the bottle of fire starter were considered in reconstructing the accident.

After extensive testing by this investigator and all of the experts on both sides, analysis of all of the documents and videotaped depositions, the conclusion of this investigator indicates that the case, as presented by the plaintiff's attorney, is highly suspect and the opinions of the plaintiff's experts are unreliable and without scientific substance.

Background to the accident

On or about February 2010, Plaintiff "A" received burn injuries at his home located in West Virginia. According to the Complaint, he was using a product called Natural Fire Starter Gel, which is an ethanol-based fuel gel.

According to testimony of "A" and his parents, a fire had been burning in the fireplace on the lower level of the house for much of the day. The Natural Fire Starter Gel was near the hearth. At some point in the evening, A was given permission by his mother to go to the lower level of the house unsupervised to roast a marshmallow prior to taking a bath He was naked,.

According to the complaint, A applied the Natural Fire Starter Gel to kindling wood in the fireplace and the Natural Fire Starter Gel container suddenly failed near the lower portion of the bottle, projecting flaming gel onto "A", resulting in injury.

Natural Fire Starter Gel is designed specifically for pellet and wood burning stoves. It is manufactured and assembled by Packaging Service Company (PSC) using a plastic bottle manufactured by C.K.S. Packaging, Inc. and a cap manufactured by Stull Technologies, Inc. The product is distributed by Hearthmark, LLC. The sole purpose of the gelled product was that the product was to be used for only wood pellet and wood burning stoves.

The subject bottle was purchased by Plaintiffs at a local Walmart store. According to the Complaint, Plaintiffs' allege the Natural Fire Starter Gel was not reasonably safe for its intended use, was defectively designed and manufactured, and carried inadequate warnings.

The undersigned reviewed over 10,000 pages of documents obtained through discovery in addition to performing, examining, or analyzing the following:

1.the subject bottle and exemplars
2. Testimony of injured party
3. View of fireplace and living room after the accident
4. Analysis of the PVC container
5. Analysis of the stopper for the PVC container
6. Proving or disproving the testimony of the plaintiff, his parents and brother
7. Testing and testimony of plaintiff's experts
8. Testing and testimony of the defense experts
9. Deficiencies in testing by defendant's experts
10. Deficiencies in testing by plaintiff's experts
11. What are the scientific basis, from a mechanical engineering stand point why either the plaintiff or defense experts were incorrect in their analysis?
12. Testimony of all experts.
13. Testing by all experts
14. Medical documents

In addition to the above, it was important to evaluate all of the chemical, physical, thermal, and flammable properties of the product in addition to evaluating the warnings and instructions.

1. **Accident Reconstruction**
2. **Annealing**
3. **Basics of blow molding**
4. **Basics o f ethanol based fires**

5. **Basics of gelled fire starters**
6. **Bottle closure**
7. **Brittle versus ductile fracture**
8. **Compatibility of a solvent and chemicals with P VC containers**
9. **Compatibility of ethanol and PVC**
10. **Consumer Product Safety Commission Reports**
11. **CT scans (Computerized Tomography Scan)**
12. **Curvature of bottle at shoulder**
13. **Deflagration, detonation, explosion**
14. **Destructive testing**
15. **Design of PVC container**
16. **Expansion o f PVC plastic**
17. **Explosion limits**
18. **Flame arrestor technology**
19. **Flame speeds**
20. **Flashback**
21. **Flashover**
22. **Heat damage**
23. **Heat exposure and PVC**
24. **Laminar flame speeds of ethanol-air mixtures**
25. **Material Safety Data Sheets (MSDS)**
26. **Napalm**
27. **Non destructive testing**
28. **Origin of fracture**
29. **Pictorials**
30. **Plastic bottle rupturing**
31. **Plasticizers**
32. **Pressure release valves**
33. **Prior reported accidents**
34. **Recalls on subject product and similar products**
35. **Scanning electron microscope**
36. **Seams in plastic container**
37. **Similar products**
38. **Single holes versus multiple holes**
39. **Stains in plastic bottle**
40. **Stress in plastic containers**
41. **Stress cracking**
42. **Stress whitening**
43. **Specification of PVC container**
44. **Supersonic versus subsonic**
45. **T ensile strength of PVC**
46. **The effect of dye in ethanol in PVC bottle**
47. **The effects of gelling ethanol**
48. **Thermal gravimetric analysis**
49. **Viscosity**
50. **Volatility**

51. Warnings and instructions

The Plaintiff, 7 year old boy "A"\ (Deposition summary with comments)

At the time of the incident, "A" was 7 years old and claims in his deposition, that he shoots his own gun. He started shooting at the age of 5. He stated in his deposition that he has never started a fire without an adult present. His mother told him not to use the gel on a burning fire because it was a waste . ***He stated his mother told him that he could start the fire that day. Contrary to that statement*** **"A"'s mother testified that she started the fire prior to taking the family to church.**

"A" stated that he ***"used the gel to start a fire and also to restart a fire."*** **This** ***admission is important and*** **more than documents the fact that "A" squirted a lot more of the Diamond product into the fire than he testified to.** ***On the day of the incident his "mom's back was out and he was sick with pneumonia."***

He stated that he was getting ready to roast marshmallows. However, his mother told him to take a bath. He asked her if he could roast a marshmallow. She said yes. He went to the kitchen, opened up the cabinet ***and got a bag of marshmallows.*** **The alleged bag of marshmallows was not located or identified in any of the documents reviewed or photographs taken at the scene of the accident.**

According to "A", the fireplace was in the basement and his mother, father and brother were upstairs. He walked down the stairs, went to the fireplace and stated that there was no fire . "A" testified that the coals were black with white powder over them, and he thought the fire had gone out. ***If there was white powder all over the coals, how could he see the color of the coals?***

He then stated that he set the marshmallows down and then put one on the stick and laid the stick right by him. He then went and obtained the fire gel. ***He moved the screen and put it on the side.*** **Moving the screen left him without any protection whatsoever. He put some wood up in a teepee and started squirting gel on it. He testified that the gel landed on a coal and started lighting. The heat emanating from the coals is over 400 degrees Fahrenheit and impossible for anyone to expose any part of their body especially an individual's hands. He then stated that he had the bottle up about a foot and the fire went up 6 or 7 inches. This was not accurate based upon the testing performed by the undersigned.**

"A" stated that he held the gel bottle about a foot from the teepee made of wood. He was on his knees andwas holding the bottle and squeezing it angled downward . He then stated thatas he tilted it up and and backward, the gel expelled out of the bottom.He stated it exploded and his brother heard the noise upstairs while he was watching television.

He then stated that he was forced backwards onto the couch and got up and ran into the bathroomon the same floor. He grabbed a cup of water and threw it on theback

of his neck in an attempt to put the fire out. He then ran upstairs to his mother. He stated, contrary to what "T" testified to, that his father grabbed him) and threw him in the bath tub and started running the water.

"A" testified that his mom and dad taught him how to start a fire using a teepee. Contrary to that statement, his father stated under oath that it was "A"'s idea to use the teepee. ***Regardless, starting that type of fire did not include using the gel.***

Although he had observed his mom's friend start a fire using lighter fluid, he has never observed any adult pour lighter fluid on a pre-existing fire.

On the day of the fire, his mother started the fire prior to going to church. The fire was burning most of the day and went out approximately 1 or 2 o'clock in the afternoon. He stated the accident occurred at approximately 8:30 p.m. If that is true and no one replenished the fire with additional logs, it would be questionable if any embers still existed. ***Note - How would*** **"A" would not be able to build a teepee if the embers were still hot** ***and not all ash?*** **He stated that there were no coals in the fireplace, there were just ashes. He also stated that matches were "right there." He also stated that if the coals were hot, he would have just put paper on the coals and not used the fire starter gel.** ***("A"'s testimony does not make sense. If he didn't use paper and he just observed ashes how could he have built a teepee if there was sufficient heat to ignite the gel? There was a reply by "A" that he did not feel any heat while he made the teepee.***

At the time he was squirting the gel into the fire, he had to stand on the ledge (hearth extension) and lean in to pour the gel onto the wood. He was able to get up on the hearth using his hands and knees. He set the gel down, climbed up on the ledge and started squirting all around the teepee. He ***stated he squirted all around before there was a fire.. He also stated that he was still on his hands and knees when the fire came at him.***

Comment: Reconstruction of the accident by the undersigned with an exemplar fireplace documents that the fire does not flash back ***but goes directly in the vertical direction.*** **Fire travels directly in the vertical direction.** ***At this time, while on his hands and knees he tilted the bottle back and the bottom came out with a noise.*** **Contrary to the analysis of the plaintiffs' experts, this more than documents that their analysis is without merit and without a scientific basis.**

The amount of squeezes the plaintiffs' expert had to complete prior to meeting within the explosive limits of the ethanol does not coincide with "A"'s testimony or the real-world.

"A" stated that when he first picked up the gel bottle, ***it was not warm. As he fell backwards, he stated that he threw the gel bottle onto the couch. (If there was any leak or opening in the bottom of the subject container, the gel would have***

squirted out the leaking end as well as the opening at the top. "A" never testified that as he was squeezing the gel into the fireplace that any other part of the container was leaking.)

He had no clothes on when his mother told him to take a bath. However, his older brother, "T", stated that his mother used the wet towel from "A"'s bath that he took prior to roasting the marshmallows in the fireplace.

Further in his testimony he stated that he used a lighter to start fires and would get gel and squirt it on the wood and light it. When he was finished lighting the fire, he would close the cap and set it down about 10 feet from the fireplace. He would then put more wood on the fire. , "A" also stated he was told to be careful of the gel. ***He knew that the gel "could light on fire" and it was dangerous.***

"A" stated that he thought the fireplace was still lit. He had observed that fireplace with the wood burning hours before that time. ***(If he thought the fireplace was still burning, why didn't "A" first try paper on the coals as he was taught?)***

He stated that he used the gel even though the ashes were still warm because he wanted to get it started. **He described the ashes as whitish, black and grey. He states that he did not see any red..** ***However, he previously testified that the coals were not warm.***

Further on in his testimony on "A" stated that the fire was out about 1 p.m. and he went down to roast marshmallows around 8:30 p.m. (the incident occurred at 7 p.m.). He did not poke a stick in the ashes to see if there were any embers. ***(His parents stated that they taught him how to start a fire and use the gel which is not consistent with "A"'s behavior and actions.)*** **"A" was allegedly taught how to start a fire and use the gel, which is an unreasonable misuse when it involves a child.**

In addition, based on "A"'s stated awareness and knowledge, and based on the fact his mother taught him not to put the gel on an existing fire or hot embers, "A"'s actions were not reasonable. Moreover, it is not reasonably foreseeable that a 7-year-old child will use the fire starter gel to start or accelerate a fire.

He reiterated the statement he made previously. "A" stated that he squirted the fire starter gel on the hot coal and it flamed up only about 6 or 7 inches. ***(However, he previously stated that the coals were not warm.)*** **As he tilted the container it came closer to the fire and then "it exploded." This statement cannot be accurate, because the air space contains close to 100% ethanol vapor. The explosive limits within the container are not reached until "A" squirts the gel into the fireplace a number of times so that air can replace the ethanol vapor and bring the concentration of the ethanol to within the explosive limits.** ***Hypothetically, assuming, for example, that there was a leak in the bottom seam, deflagration could not***

occur because the opening in the bottom would have released all pressure created by the reaction taking place.

Even though the bottom came closer to the fire, he never saw the flame come in contact with the bottle. "A" reiterated again that once he was on the hearth, he was just on his knees with his hands on the bottle. ***(Note - It would be impossible for the bottom of the container to come near the fire because "A" was squirting into the fire with the nozzle facing the fire and not the bottom of the plastic container. In addition, if "A" was squirting the bottle as he testified to, the gel would have leaked out through the small opening at the bottom if it did exist. There was no testimony by "A" that, as he was squeezing the gel into the fire, that any of the gel came onto any part of his body from the opening in the base of the container.)***

Comment - If "A" was on his knees on the hearth, there should not be any burns below his knees. It would have been impossible for him to be thrown back onto the couch. The burn diagrams from Shriners Hospital show that he is not burned a few inches above both knees and down to his toes in the front of his body except for the inside of his right calf.

The history as recorded by Shriners Hospital for Children states that "patient is a 7 year old white male with recent past history sustained on 08-28-2010. ***The patient had been roasting marshmallows inside when an accelerant for the fire caught on fire and exploded onto the patient. The patient then fell backwards into the fire before being recovered by his father."*** **The areas that were not burned were his scalp, his back and left side. There is no mention of a couch. If he fell backwards into the fire, his back would have been severely burned. It was not. However, it was possible that he fell forward into the fire. That would account for the intense burns on the front part of the body.**

The fact that "A"'s scalp and face were not burned refutes his testimony that he raised the container above his head and it exploded onto his body. If that were accurate, he would have received second and third degree burns on his scalp and his eyes and eye site would have been considerably damaged. ***(This also refutes the plaintiffs' contention that there was a small opening in the bottom of the container prior to the incident.)***

"A"'s mother, in her deposition, stated that she had problems recalling what the warnings stated on the subject container. However, she did remember that the product was flammable based on her experience. If that is true, how could she expect "A" to recall or fully understand any part of what she read and understand the meaning and intent of the warning?

"Father"

"A"'s "Father" also stated that he taught "A" at the age of 5 or 6 to make a fire. He taught him to place a piece of towel or tissue under the small kindling and light it. "A", he stated, came up with the teepee approach. ***(This is contrary to "A"'s testimony.)***

"Father" stated that he used the gel less than 10 times to start a fire and "A" watched him do it. He never saw "A" start a fire with the gel. More importantly, he told "A" that he was never to use the product because it was dangerous. Also, he never put the gel on the mantel and never saw it there. "Father" started the log fires with starter logs.

He purchased only one bottle of the Diamond product at Wal-Mart. That was the only time he ever purchased the product. This comment is contrary to "K"'s comment stating that she purchased it. He stated that he read the label and it said "keep out of the reach of children." That is why he told "A" not to use it. He does recall that it said to "store away from flame or heat sources." That is why he kept it on the floor which, in his opinion, was the coldest point in the room.

It should be noted that, contrary to the plaintiff's allegations, if it was kept on the floor, that would mean that it was well below the hearth and not be exposed to any heat from the fireplace which, the plaintiff's claim caused a leak in the container that ended up causing the fire and the deflagration.

The statement by the father also contradicts "T" who stated that the gel was kept on the mantel. "Father" always let the fire die down in the late afternoon and early evening. He again contradicted "T" when he stated that he tried to smother the fire with a towel.

When he went down to the basement, he does not remember seeing the gel bottle that day or after the fire. He did hear that "A"'s grandmother found the bottle of gel outside. I was under the couch. After the accident, his wife obtained a restraining order preventing him from seeing the children.

"Father"' stated in his deposition where, he quoted an entry in "K"'s (his wife) diary. It is alleged that after she got home on the evening of March 10 and checked the fireplace where the incident occurred, she ***discovered that "A" did not squirt the lighter fluid*** **(The gelled product)** ***into the fire as she first thought. The bottle was sitting close to the fire and appeared to have a melted crack near the bottom of the bottle allowing it to spray into the fire causing "A"'s burns. By that time, she had already retained a lawyer to sue.***

Question: "K"'s entry into her diary contradicts everyone else's testimony under oath. Her entry infers that "A" was not on his knees on the hearth holding the container as he testified to under oath. It also contradicts "Father"'s testimony. However, it does fit her attorney's theory and that of

the plaintiff's experts. This author, having over 40 years of experience in evaluating similar cases, the statements made by the family members are incredulous.

"Father" was not aware that "K", his wife at the time, allowed "A" to use the gel under her supervision. He did not expect "A" to start a fire on the evening of the incident. He also stated that before the incident, he thought he heard a popping noise. The question is whether you can actually hear a popping noise in the basement while he is on another floor watching and listening to a TV program.

"K" (mother of "A")
"K" stated n her deposition that "A" should not start fires by himself. She also stated that she never *caught him starting a fire.*

Contrary to "Father"'s testimony, she stated that *she bought the fire starter* because it was thick and it was right beside or right near the fire logs in the Wal-Mart store. <u>Without reading the warnings</u>, *she felt that the product was safe and could be used to continue the fire.* She also knew to keep it out of the reach of children and that the product could explode if it got too close to the heat source.(This contradicts "A"' testimony where he stated that his mother read the warnings to him in the store.)

It should be noted that there is not one word about explosions on the Diamond product label. **When "A" and his mother returned from Shriners Hospital in May he told her that *he was squirting the gel on the fire*. She also told aa registered nurse in the emergency department in the hospital that "*The patient squeezed the accelerant into the flames and exploded on the patient.*" At her deposition, she stated that she didn't recall telling anyone that statement. However, the statement made and recorded in the hospital notes fit the reconstruction of the cause of "A"'s burns and contradict the statements made by "A" and "K" under oath and also contradicts the opinions set forth by all of the plaintiff's experts.**

She also claimed that she saw a hole in the bottom of the bottle and assumed that it must have sat near the fire because it is only that it could have gotten there, because it got hot and melted.

She does recall seeing the marshmallow on the end of the fork .A small residue of what might have been the remains of a marshmallow was noted in one of the photographs. However, that evidence was never retained for examination. The "bag" of marshmallows were never found.

"K"'s mother found the bottle of the gel outside of the front yard under the couch. There is no explanation how her mother was able to remove the plastic container from under the couch. *It should be noted that some of the deformation on the subject container was due to the weight of part of the couch on the container caused deformation after the incident.*

She does not recall reading part of the instructions where it stated that one should "Close bottle and remove from immediate area as vapors are flammable." However, she knew that it was flammable.

When asked if she ever read the instructions on the Fire Starter Gel to "A", she replied "I do not recall." However, a question was then asked again of the reading of the label and she stated that she read it to both "A" and "T" and told "A" that he cannot use the product .

"K" stated in an answer to interrogatories that "A" had previously used the Fire Gel) under her supervision). When "K" allowed him to use it under her supervision, there was no fire. His father objected to allowing "A" to use the product. "K" does not recall how many times she allowed "A" to use the product. This contradicts her prior statements under oath. In fact, "A"'s father told "A" not to use the product.

- **"D" (registered nurse)**
  Donielle is a registered nurse working full-time for the emergency department at the Memorial Hospital and the County EMS.

Her intake states that the patient was roasting marshmallows in the fireplace. *Patient squeezed an accelerant onto flames exploded on patient.* Patient had severe burns over entire body except for his back. Estimated 70% of second and third degree burns observed. No other injuries from explosion.

This recorded statement contradicts "A"'s testimony, his mother's, his father's and "T"'s.

- **"J" (registered nurse)**
  "J" is an EMT-B. He also testified that he was told that "A" went down to the fireplace and *sprayed some type of aerosol or something on the fire and it blew back on him.*

Shriners Hospital (dictated by "R" MD in April of 1020)
The patient had been roasting marshmallows inside when an accelerant for the fire caught on fire and exploded onto the patient. The patient then fell backwards into the fire before being recovered by his father. **This *more than* documents the fact that "A" sprayed an excessive amount of the Diamond product on to the existing fire. When he fell back, he was exposed to close to 1800 °F. It should also be noted the screen was not in place when this occurred.**

Critique of Dr. "G" (Expert for plaintiff)
Dr. "G"s academic background and experience is excellent. However, his background in storage, transport and processing vessels for chemicals, gasoline and other flammable products in addition to the implementation of various flame suppression and arresting technologies for the safe handling of combustible products

in general and gasoline tank design and leak detection systems for gasoline stations and gasoline storage systems does not relate to the issues and the cause of the accident in the subject case. The "simple screen flame arrester for the inside of a gasoline tank" is not the solution in this specific case. The product in this case is thixotropic and the design of the opening in the cap must be at a size that would allow an individual to squeeze out a desired amount with minimal pressure on the container.

Dr. "G" conveniently forgets about the fact that the plaintiffs testified that "A" was going to roast marshmallows. There was no evidence found of any marshmallows in the area of the fireplace or the living room except for the small amount of white residue on the fireplace fork. There was some white color or remnant on the end of a fork in one of the photographs. However, that was never examined by any of the investigators.

He goes on to state in his report that that "A" "holding the bottle with both hands with the nozzle pointed down and pored the kindling wood when the fire gel hit the kindling/embers and then the heat source ignited the bottle due to the flashback."

Dr. "G" never attempted to reproduce the subject accident with an exemplar fireplace and the subject Diamond formulation. He accepted the questionable testimony of "A" and his mother when designing and performing his testing.

***How could* an explosion or flashback cannot occur with one squeeze when the head space contained almost 100% ethanol vapor and testing by the plaintiffs experts documented that a large number of squeezes were necessary to reduce that concentration of ethanol vapor and have sufficient enough air in the head space so that the vapors could be ignited. The experiments performed using an exemplar fireplace by the undersigned disproved all allegations made by the plaintiff and the plaintiff's experts.**

Dr. "G" neglected to state that "A" testified he was on his knees on the hearth and that he allegedly made a teepee out of kindling and squeezed the bottle on the teepee. He also neglected to state what the rest of the family testified as to what occurred as well as other recorded evidence. He also did not mention the fact that there was no evidence of any kindling in the area of the fireplace or the fire. *If "A" constructed a teepee out of kindling, where was the kindling wood located? Not one of the photographs contained any evidence of kindling wood.*

There is no evidence that the container caught on fire and exploded. If it did, it is more likely than not, that all of the contents would have been burnt up and there would have been extensive deformation of the subject container. His explanation of the deformation of the container is incorrect. The plastic container can be deformed, as it was in this case, when a weight of sufficient size is placed on the plastic for a period of time. The deformation was mechanical and had nothing to do with the incident.

More likely than not, the fire test performed by Dr. "D", another plaintiff's expert , documented that the container had to be placed in a specific position for an extended period of time before it would catch on fire and deflagrate. The amount of time that Dr. "P" had to squeeze the container before anything was notable was well over two dozen times.

Analysis, methodology and opinions (Part of Dr. "G"'s report)

A. "G" Opinion: The rupture to the subject bottle is consistent with an internal explosion.

Comment: The rupture to the subject container is not consistent with an internal explosion. ***The rupture was more consistent with the deflagration. It also required a very large number of squeezes at a specific angle over a long period of time before the fire was spread inside the container before the rupture occurred, and this was not from a "flash back" occurrence. Tests performed by the undersigned with the Diamond container filled with the thixotrophic formulation and refute the experts theory and opinions.***

B. "G" Opinion: The burns on "A" are consistent with burning gel being expelled from the bottom of the ruptured bottle and spraying on his body.

Comment: The burns on "A" are also consistent with "A" spraying an excessive amount of gel in many areas *(into the fireplace and onto his body)*. It is also consistent with "A" falling *back* into the fireplace *(see medical documentation)*. This can expose "A" up to temperatures reaching up to 1,800 degrees F. No one knows exactly how many times "A" squeezed the plastic container. *For example, it is obvious that he squeezed a large amount of the gel directly in front of the hearth. There is also evidence that there were many squirts all around the fireplace, on areas surrounding the fireplace, the walls and ceilings.* As proven by the undersigned, in all of the experiments performed, no flashback occurs at anytime no matter how much gel is squirted into an open flame.

C. "G" Opinion: Explosion tests were conducted with 5 exemplar bottles of Natural Fire Starter Gel. Ten tests were conducted with single hole caps by squeezing the bottles of gel over a candle flame. In each of these tests, the bottle had an internal explosion. In one of these tests, the bottle ruptured. The ambient temperatures of these tests were between 55 and 65 °F.

Comment: The explosion tests that Dr. "G" conducted on five exemplar bottles did not involve methodology that was scientific. The tests conducted on 8/29 self extinguished. No explosion occurred. The container deflagrated.

The test conducted on 9/11 by "G" where the base of the container blew out did not, in anyway, reproduce the results produced in the exemplar container. The container was held in a horizontal position for an extended period of time which is unrealistic. The same is for the second test that was held on 9/11.. In addition, the initial ignition of the external vapor was snuffed out because the internal concentration of the ethanol was not within the explosive limits.

There is no documentation that the subject container was only partially filled containing a lot of head space. Flammability tests performed by the undersigned using an exemplar fireplace disproved the theory and opinions presented by Dr. "G". The thixotropic Diamond product was squirted directly in front of the fire exposing the spout, vapor and thixotropic liquid to over 200 degrees Fahrenheit and through a hole on the top of the fireplace exposing the spout, vapor and thixotropic liquid to over 300 degrees Fahrenheit. No flashback or internal deflagration occurred while the undersigned personally squeezed the plastic container. The spout was so hot that the undersigned was not able to close it until it cooled down

The additional test *performed by Dr. "G"* with a different cap are also non scientific and do not, in any way simulate the conditions that existed at the time "A" squeezed and squirted the gel all over the pre-existing fire, carpeting and walls.

Using the same formulation with a different cap documented what probably occurred on the original cap on the subject bottle. The subject bottle did not explode but the squirting of the gel all over the fireplace by "A", including his body on a pre-existing fire significantly contributed to "A"'s burns when he was ignited.

If "A" was really going to toast marshmallows, there would have been evidence of marshmallows that were roasted on the pre-existing fire and glowing embers. There was no evidence in the room where the fireplace was located that there was even a supply of marshmallows anywhere else in the house. "A" nor his mother never testified where the marshmallows were originally located. "A" never testified that he took the fireplace fork into any other part of the house and placed a marshmallow or marshmallows on the fork.

D. "G" Opinion: The concentration of the vapor space in the gel bottle was measured with an O2 sensor. This concentration was found to be in the explosive range for temperatures above 50 °F.

Comment: When the cap is first removed the headspace contains well over 19% ethanol vapor. When the container is constantly squeezed, external air constantly dilutes the concentration of ethanol vapor in the headspace to a point where it can fall within the explosive range of 3 to 19%. This can only occur if one squeezes the container a large number of times. That is the only way that the plaintiff's experts were able to create sufficient oxygen in the headspace when the container was constantly squeezed and held in a horizontal angle. When the concentration of ethanol vapor is above 19%, the flame front will be snuffed out.

E. "G" Opinion: The cap on the Natural Fire Starter Gel bottle had on 0.113 inches (2.87 mm) hole and with this cap all of the tests showed internal combustion.

Comment: The statement is not true. Most of the ignition tests were self extinguished *because the concentration of the ethanol vapor was above 19%.* It is only when the container was continually held in the horizontal position on a continuous basis where air was introduced into the air space and the concentration of the ethanol vapor falls within the explosive limits did Dr. "G" obtain internal combustion and not an explosion. If he recalls, "A" testified that he was holding the container in two hands and aiming downward. If that is the case, then there would be no head space and no internal combustion. In the tests that Dr. "G" conducted he should have directed the cap in the downward direction rather than horizontal. In addition, had he placed more gel in the container, it is more likely than not it would have no internal combustion taken place at all.

F. "G" Opinion: Four tests were performed with Kingsford charcoal lighter fluid caps (3 smaller holes) on a Natural Fire Starter Gel bottle. There were no internal combustion for these tests.

Comment: The tests conducted with three small holes were not conclusive and the methodology was not scientific. In addition, the opening in the subject cap was designed to expel a thixotropic formulation. Three holes could have been used for a liquid but not the thixotropic formulation. The use of a three-hole cap would not prevent the underlying incident and would not have acted as a flame arrestor. Moreover, a three-hole cap could expose users to more dangers because if they misuse the product near a flame or hot ember, the consumer has to position the bottle closer than the one with the one-hole cap. In addition, the end of the cap was held much further away from the candle than all of the prior tests with the standard closure.

G. **"G" Opinion: The Consumer Product safety committee induced Anderson Chemical to recall 168,000 bottles of Eco-Lite Charcoal starter in 1991. The CPSC learned of five reports of the bottles exploding in consumers' hands when this charcoal lighter fluid was sprayed on hot charcoal. Four of these explosions involved serious injuries, including hospitalizations for burns.**

Comment: The recall of the Eco-Lite Charcoal Starter in 1991 had nothing to do with the subject formulation. That product was a liquid formulation not a thixotropic and was designed for a different intended use than the product at issue. A red herring.

H. **"G" Opinion: Eco-Lite (82% ethanol, 8% isopranol, 3.5% methanol) is made of a mixture of alcohols and has a flash point of 56 °F. The other more familiar charcoal lighter fluids are made of petroleum products and have a flash point above 100 °F. It should be noted that the Diamond Natural Fire Starter Gel (85-95% ethanol) is made of ethanol and Bitrix and it has a comparable flashpoint to the Eco-Lite mixture.**

Comment: What Dr. "G" does not understand is whether the flashpoint was 56 °F or 150 °F, it wouldn't make any difference when the vapor is exposed to an ignition source of 1,500 to 1,800 °F. In addition, the flash point reported was incorrect. The closed cup flash point was significantly higher.

I. **"G" Opinion: Caps with 3 small holes are commonly used in the industry and well known. A standard 3 hole cap, manufactured by Stull Technologies, would have prevented this incident as the smaller holes act as a flame arrestor. An igniting flame cannot pass through the small holes and into the container to cause internal combustion. Thus, an explosion cannot occur.**

Comment: Dr. "G", in all his tests, did not reproduce the alleged blow (deflagration) out of the subject container in any of his tests. His statement regarding flame arresters is utter nonsense.

The undersigned has been involved with exploding batteries with micro porous flame arresters for over 30 years. The flame front goes through the flame arrester into ullage area much too fast for even the plastic to melt and snuff out the flame front before the battery explodes. The purpose of the flame arrester is to snuff out the flame before it reaches the internal head space of the flammable liquid. When it comes to polymeric materials, the flame front does not melt the opening fast enough to snuff out the flame. It is not foreseeable when one performs a risk analysis on the subject formulation that a foreseeable consumer would constantly squeeze the container when only a half a tablespoon would be used to start a fire. The product was not designed for a 7 year old to use the product as a toy. That is not a foreseeable misuse of the Diamond product.

J. "G" Opinion: Warnings do not alleviate the product manufacturer and retailer of the duty to design and sell a safe product. It is a basic, generally accepted mechanical engineering principle that performance of a risk assessment (e.g. failure modes and effects analysis) is necessary in product design. It is also a basic, generally accepted mechanical engineering principle to preferentially design out risks as opposed to warn against them. Manufacturers, distributors and resellers have a duty to consider anticipatable uses and misuses of a product.

Comment: The warnings and instructions on the subject container were well done for a pictorial of a wood pellet stove. *The product is not designed for 7 year olds. It is designed for a small segment of our population of adults who own wood pellet stoves.* It is not designed for fireplaces otherwise a pictorial of a fireplace would have been on the front panel of the product. The wording on the labeling on the container covers that specific area.

Dr. "G" fails to point out that there were starter logs on the side of the fireplace and that the parents knew that there was a safe method of starting a log fire in their fireplace. He also fails to point out that the parents knew that the subject fire starter was highly flammable and dangerous, must be kept away from children and was designed for wood pellet stoves as shown on the pictorial on the label of the product. The warnings were written for a person with a fifth grade education and would be obvious to any consumer that the small container, like gasoline, should be kept in the garage and out of reach of young children. A manufacturer is not responsible to follow a consumer to their residence and hold their hand while they use their product.

K. "G" Opinion: It is apparent that none of the defendant's *experts* in this case performed any tests of any kind to determine if the container and cap could be safely used by a consumer in the presence of a flame. Had the management or engineers for these defendants conducted appropriate testing of this product, they would have determined that the Diamond Natural Fire Starter Gel and its container pose a significant and unsafe explosion risk.

Comment: The defendants did not have to perform any tests because the properties of the gel are known as well as the use and application for pelletized wood stoves. The properties of the one hole cap and its integrity and use for thixotropic formulations is well known for many applications including harmless products such as soap. The same design is also used for squirting insect repelling products inside and outside residential and commercial buildings.

The public also knows that gasoline in any container poses a significant and unsafe explosion risk. The same goes for Drano which is composed of almost 100% of sulfuric acid. *These* products are not designed for 7 year olds.

What Dr. "G" fails to point out is that there have been no reported accidents with fireplaces and the subject product with young children. In addition, there have been no reported accidents with wood pellet heating stoves especially when only a tablespoon of the Diamond product was used to start the fire in a wood or wood pellet stove.

This author measured the force required to release the snap cap of the subject bottle. The tests indicate that it takes over 8-pounds of force to pull the cap off. The cap would be considered child proof. Therefore, it is questionable whether a 7-year opld would be able to easily open the cap with one finger.

L. "G" Opinion: It is also apparent that none of the defendants in this case performed a reasonable risk assessment of this product kind to determine if the container and cap could be safely used by a consumer in the presence of a flame. Had the management or engineers for these defendants conducted an appropriate risk assessment of this product, they would have determined that the Diamond Natural Fire Starter Gel, its cap, and its container pose a significant and unsafe explosion risk. Further they would have discovered the problem had been solved centuries ago and the solution implemented.

Comment: The problem that was allegedly solved centuries ago did not involve 7 year olds. The flammable tests using an exemplar fireplace performed by the undersigned refute all of Dr. "G"'s opinions. In addition, he is not in any position to come to that conclusion without performing his own tests with an exemplar fireplace and the subject Diamond formulation.

M. "G" Opinion: This product as packaged is unreasonably dangerous to be sold as a consumer product.

Comment: The product as sold to the consumer was not unreasonably dangerous because it was not designed or sold with the intent that a 7 year old would be squirting the contents of the bottle on a fire. *In fact, there is no evidence that "A" squirted the contents of the bottle on embers. There is evidence to the contrary that he actually squirted the contents of the bottle onto the pre-existing fire.* In addition, no one knows how many times or how much of the gel that "A" squirted onto the fire once it

got started. He also squirted additional amounts to restart the fire. ***We also know that he squirted the liquid in a large number of other areas. This is illustrative of the fact that "A" was playing with the Diamond product and did not use any of the warnings and instructions his mother stated she gave him previously.***

N. "G" Opinion: An internal explosion can occur in a Diamond Natural Fire Starter Gel bottle as it is being squeezed onto a flame or ember.

The statement that Dr. "G" made has a low probability ***and is totally false.*** **Experiments performed by the undersigned continuously documented that no explosion or deflagration can occur internally no matter how much gel was expelled into the fire from the front or directly on top of the fire. This was documented in the videotape test performed by the undersigned. It is more probable than not that "A" had to squeeze excessive amounts of the gel on the fire in front of him.** ***What no one can answer at this point is when the bottom seam separated and melted. It is more probable than not that the bottom separated well after "A" was on fire.***

In addition, not one of the experts for the plaintiff was able to reproduce the appearance of the subject container in anyway whatsoever. Similarities resulted by squirting the container excessively until sufficient air entered into the head space and reduced the concentration of the ethanol vapor to within the explosive limits. ***This illustration by the plaintiff's experts did not, in any way, simulate "A"'s testimony.***

Squirting the contents excessively into a burning fireplace caused a large flashback onto "A"'s exposed body, ***which had been previously squirted on by "A" himself,*** **to the elevated temperatures emanating from the fireplace while he was** ***allegedly*** **on his knees facing the unscreened fire.** ***The high temperatures emanating from the fireplace as measured by the undersigned in the tests performed, was more likely than not, much too high for anyone to withstand.***

Contrary to Dr. "G"'s statement, the subject product was not designed to be squeezed on glowing embers or pre-existing fire ***in a fireplace*** **on a continuous basis by an adult or by a 7 year old.** ***The product was created and sold for pellet and wood***

stoves only. This was open and obvious to any adult looking at the label on the container. **If several tablespoons were placed on the glowing embers or pre-existing fire, "A" would not have been injured.**

O. "G" Opinion: This internal explosion can occur when the ambient temperature is above 50 °F.

Comment: Dr. "G"'s tests were not supported by scientific testing and documents readily available in the subject case.

P. "G" Opinion: While the outside temperature was 24 °F, the accident happened in a finished basement of a house where a fire had already been burning to roast marshmallows so the temperature at the time of the incident was likely over 50 °F.

Comment: ***There was no testimony or comment by "A" that, prior to the subject incident, he was able to safely roast and consume a marshmallow.*** **Not one of the experts reported marshmallows being roasted or present in the area. If the experts continually believe the mother and "A", they will continue to make errors and produce results that are not scientifically acceptable. In addition, there was no evidence of a bag of marshmallows available anywhere.**

Q. "G" Opinion: The 5 previous explosions of the Eco-Lite Charcoal starter in 1991 are very similar to the explosion of the present case. The flash point for this previous Eco-Lite product and the present gel are both approximately 56 °F and the chemical compositions of the two products are substantially similar.

Comment: The formulation for the Eco-lite Charcoal Starter in 1991 was a pourable formulation and is not related to the gelled thixotropic formulation. These two formulations should not be intermixed.

R. "G" Opinion: The flashpoint for alcohol is particularly hazardous since at room temperatures, alcohol vapors are in the explosive range. Petroleum distillates like diesel fuel have a higher flashpoint and their vapors tend to be too lean at room temperatures to be explosive. Fuels like gasoline with a very low flashpoint have vapors that tend to be too rich at room temperatures to be explosive.

Comment: When alcohol is placed in a thixotropic formulation, it is much less

dangerous than the pourable liquid. The tests performed by the experts document this statement. Furthermore, the flammable tests performed by the undersigned using an exemplar fireplace and the thixotrophic formulation demonstrated the integrity and the formulation of the container.

S. "G" Opinion: A flame arrestor could prevent internal explosions of this alcohol gel. Flame arrestors have been proven to prevent the propagation of flame into a container. Even a cap with smaller holes, such as the Kingsford cap tested by BEAR, can act as a flame arrestor.

Comment: Dr. "G" failed to test the squirting capabilities of the smaller holes versus the one in the subject container. What he also is demonstrating is that the subject gel does not flow in comparison to a non-gelled liquid. If the product is so highly flammable, it would have emitted sufficient vapors and ignited when first exposed to the radiated temperature emanating from the fireplace.

What Dr. "G" fails to demonstrate is that the open holes have to be small enough to snuff out the flame front entering through the hole openings in the cap, or that the openings will melt and close fast enough to snuff out the flame front before it enters the head space area in the container. That is impossible.

In summary, I am sure that the billing for Dr. "G"'s work conformed to the inadequate testing over a long period of time.

There were many areas that proved that the subject product, when used in accordance with the pictorial that was on the front of the label was relatively safe and not designed to be used by 7 year olds who were suppose to be roasting marshmallows.

More probable than not, based upon over 50 years of experience in the area of flammable materials, fires and the reconstruction of accidents, "A" did a lot more than just squirt the thixotropic product on the fire. There is no doubt that he squirted the product into a number of areas ranging from the carpeting near the hearth, the hearth, the walls and other areas of the living room in addition to squirting a large amount onto a pre-existing fire. The pre-existing fire was a result of "A" starting the fire and then adding the thixotropic gel in large quantities in the area around the fire, the walls, the back of the fireplace and over parts of his body. Dr. "G" ignored the reality of what actually occurred.

Critique "D" (Another PhD plaintiff's expert)

"P" describes the accident stating that ""A" Nelson, a 7 year old boy, was attempting to ignite a fire in a home fireplace when the bottle of fire gel ruptured causing the napalm gel contained in the bottle to spray back onto him resulting in severe burns."

A. **"P" states "that the bottle was an unsafe container for a napalm or napalm-like gel."**

Comment - Napalm (naphthenic and palmitic acids) is a thickening/gelling agent generally mixed with gasoline or a similar fuel used in military operations. It is a tactical weapon to remove vegetative cover. ***The product at issue is not considered a napalm gel.***

"P"'s conclusion that "the bottle was unsafe" has no scientific or technical basis."

B. **"P" stated "that the bottle was manufactured which is not *fully* compatible with the hazardous contents (i.e., ethanol based napalm gel)."**

Comment - PVC containers have been safely used for ethanol and many other solvent and flammable products for generations. The use of the word "fully", does not, in any way indicate that the contents of the subject bottle are not safely contained for its foreseeable use and application. At room temperature, the container is completely compatible with ethanol if one performs a risk analysis of the foreseeable use and foreseeable misuse of the subject product, the container will not, foreseeably, be exposed to elevated temperatures. The undersigned has personally used PVC plastic containers for ethanol from 1955 to the present time. All of the research performed by the undersigned in graduate school that incorporated pure ethanol was always transferred from large containers from the supplier to small PVC containers. In undergraduate school, the undersigned worked in the chemical supply room for 4 years transferring 80 to 100% ethanol from large containers to a small PVC containers for students to be used for laboratory experiments. The undersigned had the same experience in industry. There was never a question as to the integrity of PVC and its use as a container for ethanol liquid.

Furthermore, this investigator had, in the 90's patented an ethanol based product in a PVC bottle that contained a non-toxic repellent. Not one accident in the years it was sold throughout the United stated, Europe and China.

C. **"P" states "that the bottle wall did not have sufficient wall strength to accommodate the internal forces of an accidental internal vapor ignition."**

Comment - What "P" does not admit was that there is no polymer container that can withstand an internal the deflagration or explosion. Why would he single out this specific polymer and not include every other type of polymer? The flammability experiments performed by the undersigned with an exemplar fireplace and exemplar Diamond containers of the thixotropic formulation refute "P"'s statement.

D. "P" states "that the bottle wall had too broad a wall thickness variation leading to stress concentration in the thin areas near the bottom."

Comment - As stated previously, there is no polymer container that can be designed with respect to the subject and similar ones that can withstand internal deflagration and/explosion. However, the flammability experiments performed by the undersigned using an exemplar fireplace and exemplar Diamond containers of the thixotropic formulation refute "P"'s statement.

E. "P" states "that the thinnest area (weakest and most likely point of rupture) in the wall was near the bottom portion of the bottle which was the area closest to the person holding the bottle."

Comment - What "P" fails to state or consider is that the head space above the ethanol gel contains close to 100% ethanol vapor or, in the alternative, well over the upper explosive limit of 19%. When the container is turned upside down and a small portion is poured out (a spoonful or two), no change takes place because the thixotropic material moves towards the spout and the head space moves to the rear or bottom of the container. It is only when an individual, such as "A", continually squeezes and squirts the contents many times causing air to then mix with the head space and reduce the concentration of alcohol vapor to within the explosive limits can the vapor ignite and eventually deflagrate internally. ***This was proven and well documented by "P"'s tests. (See reply in I.)***

Furthermore, the differencews in thickness amounted the the thickness of an average human hair.

F. "P" states "that the appearance of the subject bottle is consistent with the testimony of "A" concerning how the explosion occurred."

Comment - "P"'s reliance on "A"'s deposition testimony versus everyone else's documented records is incredulous. ***He never attempted to reproduce the subject incident using an exemplar fireplace with the actual product.*** **His testing does not in any way reproduce even "A"'s testimony assuming even part of what he testified to was true. His testimony contradicts the scientific analysis of the ignition and the injuries documented.**

G. "P" states "there is no way to design and manufacture a squeezable blow-molded plastic container that can safely contain in the detonation of a napalm gel. However, it is possible to design a bottle to handle a deflagration/slow burn without blowing napalm back onto the consumer using the product."

Comment - The subject container and similar thixotropic products have been safely used by tens of millions of consumers for decades. He continually forgets that the

product was not in any way designed for 7 year olds to play with *or a fireplace*. The product was designed for a small segment of our population that used wood pellet or wood stoves.

Contrary to "P"'s opinion, if there was a practical and inexpensive design for the same container to handle a deflagration, slow burn without blowing "napalm" back onto the consumer it would have been produced and used in lieu of the subject one. If CKS was responsible, and they are not, in manufacturing a plastic container filled with thixotropic ethanol, that would be safe for a 7 year old, it would be supplied in a steel safe with no combination to open it.

If "P" performed a risk analysis relating to the foreseeable use and the foreseeable misuse of the subject product, he would be enlightened to know that the product was designed specifically for pelletized wood burning stoves and not fireplaces. He would also agree that the instructions state that only a tablespoon of the material must be placed within the pelletized wood stove burn pot, that an additional half tablespoon of gel is to be placed on top of the wood pellets, and that the product should not be squirted by a 7 year old all over a room including the fireplace where an excessive amount was squirted continually by "A". In fact, "P" reproduced how many times one has to squeeze the subject container before enough oxygen was drawn into the head space to bring the ethanol vapor concentration down to within the explosive limits. *That is not a foreseeable misuse of the product. It is not foreseeable that parents would allow a 7 year old child to use the product as a toy.*

H. "P" states "Based upon my experience I believe that bottle manufacturers should communicate with their customers to learn the intended use of their bottle. The bottle manufacturer and/or the customer should test the bottle to determine if the bottle is safe for the intended use. If the bottle is found to be unsafe for the intended use, the bottle manufacturer should refuse to market the bottle for that application."

Comment - This more than documents the fact that "P"'s experience is limited in reference to the real world of involving the manufacturing of plastic containers. The sealed area is usually the weakest part of both plastic and glass containers. "P" should also know that if he was capable of performing a risk analysis on the subject container, it would have been easy to *prove that it was not foreseeable that a 7 year old would be playing with the container as a toy*; holding it in a horizontal position adjacent to a flame for an extended period of time and then raising it above his head where he testified it exploded. It should be noted that there is not one other reported incident with the subject product where the incident was described similar to "A"'s.

I. "P" states "a PVC bottle is not appropriate as a container for napalm gel but manufacturing the bottle with the bottom being the weakest point (closest to the user) in the container was exceptionally reckless."

Comment - "P"'s comments again demonstrates his lack of expertise in the area of plastic containers. The seals of all plastic containers are the weakest part of the container. It is also the weakest part of a metal and glass containers. The subject container had seals at the base and on each side of the container. The seal that will be based on how the container was being held while the internal vapors inside the container was on fire. The seals could have failed at any of locations where the seals were located.

The subject plastic container in this case was designed to satisfy load drop-test and also satisfy the internal pressure produced by ethanol vapor. A drop test, in this connection, tests the ability of the bottle to withstand impact forces when the bottle is dropped on its mouth or neck portion because the mouth and/or neck portions are generally dimensionally the smallest parts of the bottle, any impact forces applied to those regions create maximum stresses in the walls of the container that tend to cause the container to burst or rupture at the weakest areas, typically the mold seams. Existing bottles, such as the one in the subject case, used by the chemical industry, have satisfied both requirements or specifications established by the various government authorities and transportation laws. It is important to note that a spherical surface where the seams are located in the subject container is the optimum design for the most resistance to failure.

PVC (Polyvinyl Chloride) is a softer plastic, which is relatively inexpensive. It is very durable and can be dropped without much of a problem. It also takes a lot of deliberate strength to tear PVC apart. The variation of the hardness of PVC is based upon how much of the plastic is put into a specific area. PVC can also be recycled. In addition, PVC has been used to contain ethanol and other alcohol for decades.

"P" also states in his opinion, that PVC is "marginally compatible" with ethanol and that is not true. The rest of his opinion on page 12 of his 28 page report is purely subjective without scientific documentation.

"P" also states that when he was with Dow Chemical, Dow was obligated to work with their customers to insure the safety of the consumer. In this case, CKS was not responsible to perform any risk analysis studies because they supplied the PVC containers as requested. However, he neglects to state that liquid ethanol and other alcohols such as isopropanol have been sold in PVC containers for decades. If that is true, there is no other obligation or responsibility on the part of CKS with reference to the end use application of the PVC container in this case.

If one examines "P"'s litigation experience on page 16 and his scientific presentations listed on pages 17-26 as well as the patents listed on pages 28-29, there isn't one reference that relates to PVC, ethanol, fires, explosions and deflagration of containers.

Lastly, the undersigned assumes by choice, that "P" chose not to address the subject of marshmallows even though he had an opportunity to review the depositions of "A" and other members of the family as well as an opportunity to examine all of the photographs associated with the accident. He also failed to comment on the fact that if there was an opening at the base of the container that any squeezing of the container would have released the thixotropic ethanol through the opening in the top closure as well as the alleged opening in the base of the container.

Furthermore, tests performed by the undersigned using an exemplar fireplace and exposing a half filled Diamond product container to well over 200 degrees Fahrenheit documented that the bottom seam would not rupture or melt and that the container actually expanded as shown in the videotapes and photographs taken by the undersigned during and after the tests.

Critique by "H" (Expert for the plaintiff)

Dr. "H" stated in her report that "The product is marketed as a fire-starting accessory, it is suggested for use with wood or pellet stoves."

From Section 3.0 (Background) to page 7 of Dr. "H"'s report, she re-iterates what is already known through prior publications and/or company literature. What is important on page 8 is that she measured head space of the volatiles in the PVC container and found that it was essentially ethanol. There was no report that air and/or oxygen was present in the head space. That is important because no explosion can occur if the concentration of the ethanol vapor is greater than the upper explosion limit.

Dr. "H" refers to the CPSC's 1991 voluntary recall of Eco-Lite charcoal starter. This is a product composed of ethanol and individuals were warned not to spray the material over a burning fire and/or charcoal because, like gasoline it may result in a flashback.

Comment: The objective of the Natural Fire Starter Gel was exactly what the name refers. The name and pictorial on the label designate that the product is designed to be used for wood and pellet stoves and not fireplaces. The warnings Eco-Lite product was not thixotropic and designed to start coal fires for barbeques, as an example. The two products are not designed for the same application.

It is open and obvious to a consumer that one does not squirt gasoline into any type of fire. Therefore, when the consumer is properly warned to keep the product away from children and a parent, as in this case, does not place the subject product out of the reach of their children, including "A", and claims they allow "A" to use the product. Manufacturers cannot be responsible for their actions when they violate the manufactures instructions. Manufacturers cannot be responsible for the parents'

actions when they decide, on their own, to use a product designed for wood and pellet stoves to be used in a fireplace. Dr. "H" ignores that issue.

When a product involves a fire starter of any type, there are no liquid or gelled fire starters that are inherently safe. However, in the subject case, even "A"'s father stated that he used the Safe Fire Starter logs that were on the side of the fireplace. Why any parent would use any other product when they have a safe way of starting a fire is questionable.

What Dr. "H" fails to also point out is that all of the other products do not designate that their products should be used for wood and pellet stoves with a pictorial on the front. *In the opinion of the undersigned, that significantly differentiates the subject product in this case versus all of the others.* **The CPSC staff are not warning experts and never addressed that issue with reference to the fact that the Natural Fire Starter Gel with the pictorial and the fact that it was designed only for wood and pellet stoves distinguishes that product from all of the others.**

Dr. "H" continually relates the Natural Fire Starter Gel to the Eco-Start Natural Lighter Fuel. They are two separate products designed for two separate objectives. She continually fails to recognize or even discuss the differences in the two and the pictorial of one product versus the other. In fact, on page 12 of her report, she does not feel that there are any differences in the two.

Dr. "H" relates the two by stating that they would act identically with a hot coal or a piece of burning wood "regardless of whether the products are used in wood stoves, barbeque grills, or fireplaces." Dr. "H" is not a warning specialist and, from that stand point alone, ***there is a significant difference.***

The Natural Fire Starter Gel is specifically designed for wood stoves and pellets and not barbeque grills or fireplaces. In addition, from a risk analysis standpoint, it is not foreseeable that a parent would use a product designed for wood and pellet stoves or fireplaces when they already had purchased and are experienced in using the Safe Starter Logs. Furthermore, it is not foreseeable that a parent, under supervision or not, would allow a 7 year old to use the product with or without their permission.

"H"'s reports her comments relating to the determination of the subject fuel gel versus what was reported by the manufacturer and the CPSC may vary slightly but it is not an issue in this case. The temperature of the flame of a fire is in the range of 1800 °F and that of burning coals is over 900 °F. What difference would it make whether the flash point of the gel is 54, 73 or 130 degrees? Once the vapor is ignited in the fireplace, it would still ignite and flash back whether the flash point is 52 or 130 degrees.

Therefore, the comments Dr. "H"'s report do not make any sense, and in the opinion of the undersigned, ,the fire starter is safe when it is used in accordance with its instructions. for wood and pellet stoves.

Every consumer has the right to purchase any product they want in Wal-Mart or any other store. For example, many stores sell tanks filled with propane. Propane is highly explosive and has explosive limits between 2.1 and 10.1 percent. Gasoline is also explosive and has explosive limits between 1.4 and 7.6 percent.

When gasoline is used for gasoline lawn mowers, gasoline hedge cutters and emergency electrical power generator, the gasoline is usually dispersed from small container ranging from 1 to 10 gallons. There has never been a recall for any of those products even though they are more dangerous than the thixotropic ethyl alcohol. When the thixotropic ethanol, propane gas and gasoline are used for their intended purpose by adults, the risk of foreseeable injuries is reduced significantly. Dr. "H" fails to continuously point out that all of these products are not designed for parents allowing 7 year olds to use the product under any conditions, *especially when "A" used the product as a toy.*

Lastly, Dr. "H" spent a lot a time reproducing data that is readily available by the manufacturers of the products and documents readily available on line. There was no need to perform any laboratory tests proving what is already known and attempting to reinvent another wheel.

1 "W"J.D.(Expert for the plaintiff)

Mr. "W" defines his area of product safety management as a "system that a reasonable prudent manufacture puts in place before the first product is conceived and ensures that the final product along with its warnings, packaging and marketing materials is reasonably safe."

What Mr. "W" neglected to discuss is that every consumer has the right to purchase any product they want in Wal-Mart or any other store. For example, many stores sell tanks filled with propane. Propane is highly explosive and has explosive limits between 2.1 and 10.1 percent. Gasoline is also explosive and has explosive limits between 1.4 and 7.6 percent.

When gasoline is used for gasoline lawn mowers, gasoline hedge cutters and emergency electrical power generator, the gasoline is usually dispersed from small container ranging from 1 to 10 gallons. There has never been a recall for any of those products even though they are more dangerous than the thixotropic ethyl alcohol. When the thixotropic ethanol, propane gas and gasoline are used for their intended purpose by adults, the risk of foreseeable injuries is reduced significantly. Mr. "W" fails to point out that all of these products are not designed for parents allowing 7 year olds to use the product under any conditions.

The hazards were identified in each of those products; the risks were assessed and adequate safety measures were applied to minimize both hazards. The same approach was made by the manufacturers and distributors of the Natural Fire Starter Gel which is much less dangerous than propone and gasoline. *As stated previously,*

the undersigned has been involved with the use and application of 80 to 100% ethanol and PVC containers for close to 58 years.

It should be noted that Mr. "W" is not a warning specialist especially in the area of hazardous products and chemicals. He points out on page 7 of his report that he is developed on-products warnings and instructions for a number of manufacturers and distributors. Not one of them involved flammable chemicals with physical and chemical characteristics similar to the Natural Fire Starter Gel.

Vending machines, ATVs and pool products are not at the level of warnings and instructions required for products such as the Natural Fire Starter Gel. In contrast, the undersigned has created warnings and instructions for a number of corporations, presented OSHA safety meetings at a large number of construction sites, created Material Safety Data Sheets for flammable products, evaluated polymers from room temperature to hyperthermal temperatures, consulted to NIOSH (the research arm of OSHA) and performed many explosive tests on a variety of flammable products.

Mr. "W" also claims that he created warnings and instructions for fitness equipment in Dick's Sporting Goods. The undersigned has evaluated many accident cases involving fitness equipment. I have also invented sports products that were sold in Dick's. Contrary to Mr. Kitzes' claims, the warnings and instructions for fitness equipment are created by the manufacturers of the equipment and not the management of Dick's Sporting Goods.

Mr. "W"s' report he points out the use of a pictogram but fails to point out the pictorial of a wood stove on Natural Fire Starter Gels label versus that of a fireplace. He also fails to point out that the words in the warning are at the level of a fifth-sixth grader. In addition, he failed to point out that the product must be kept away from children and is not designed to be used by a 7 year old.

The label of the Diamond product states in block letters "PERFECT FOR ALL WOOD AND PELLET STOVES." There is also a pictorial of a wood stove - NO FIREPLACE is shown or mentioned on the front or back labels of the Diamond product. Mr. "W" failed to address that issue which would be considered self-serving on his part by not commenting on that information on the label.

Mr. "W" describes "A"'s behavior prior to the accident without noting all of the different statements made to the EMTs and medical specialists at the hospital. He even failed to mention "A"'s brother's statement under oath where he stated that "A" had already taken a bath and his wet towel was upstairs in the bathroom. In fact, the description of the accident described by Mr. "W" is totally inaccurate and technically incorrect.

Mr. "W" makes a statement regarding "A" being an "avid outdoorsman, hunting, fishing and trapping." What he doesn't point out that "A" at the age of 7 was able to read and did well in school and was sophisticated for his age. ***is immature and***

irresponsible, based upon his behavior prior to and up to the time of the incident. In addition, contrary to Mr. "W"'s statement, it should be obvious to him that "K" did not read the Diamond Natural Fire Starter Gel label to "A" on prior occasions. That conclusion is obvious to the undersigned.

Although Mr. "W" had been associated in some way with the CPSC (not true), he is not a chemist, chemical engineer, safety specialist, safety engineer, professional engineer or certified in warnings and instructions. In addition, he is not qualified to perform a risk analysis, based on his background and experience on the product, Diamond Natural Fire Starter Gel.

The fact that "K", "A"'s mother, *could not read certain important parts of the label and, on top of that,* disregarded the warnings concerning children, negates all of his opinions in regarding the subject warning. When you have a consumer such as "K" dismissing the dangers which she was aware of and the warnings concerning her own child, how could Mr. "W" criticize anything on the label of the subject bottle.

Mr. "W" knew, or should have known, especially with his law degree, that when you have a recalcitrant worker on a construction job, who refuses to wear protective equipment and is injured, that worker is responsible for his own injuries. In this case, "K" openly refused, in spite of her husband's opinion, to follow the simple instructions and warning on the back label of the Diamond Natural Fire Starter Gel. Now that the horse has run out into the pasture, both Mr. "W" and "K" want to place a lock on the barn door. The undersigned handles similar situations like this each week when the son or daughter of an individual receives a brain injury in soccer and can't go back and play unless they wear his headgear (see www.forcefieldheadbands.com). By that time, it is much too late *to help the child reduce the risk of a brain injury.*

Although the warnings and instructions on a subject label were accurate based upon the manufacturer's intended use of the product in wood and wood pellet stoves, the instructions regarding children were completely disregarded by "K". If she just obeyed that one warning, "A" would never have been injured.

In spite of "K"'s behavior, Mr. "W" continues to quote various sections of the CPSC without addressing the real issues in the case. For example, on page 12 of his report he describes the vapor "may burn with near invisible flame." What does that mean to a 7 year old? The fact is that the warnings and instructions were never designed for a 7 year old to read and use the product. What Mr. "W" fails to recognize is that "K", the mother of "A", had complete control of the product and completely ignored the warnings.

There was nothing complex about the simple warning concerning children. There is nothing that a manufacturer can do if a consumer, such as "K" refuses to follow and ignores important instructions concerning the safety and welfare of her own child.

On top of all of the instructions that should have been followed if "K" used the product, no one used a tablespoon of the product at any time. If a tablespoon amount was placed on any fire, no injuries would have occurred.

The undersigned has been involved with warnings and instructions regarding all types of chemicals from the early fifties to the present time, from the warnings and instructions published by the Manufacturing Chemical Association to the current ANSI Z 129.1 which is the labeling standard for Hazardous Industrial Chemicals-Precautionary Labeling. It is important to note that this is a voluntary standard and not statutory. All of the other references Mr. "W" makes from do not relate to the real issues in the subject case.

Most important, the manufacturers of the product knew that the product will meet the needs of a small segment of the market which is wood and wood pellet stoves.

"W" states that "the highly flammable ethanol based gel with a nearly invisible flame, could flashback to the bottle, causing it to explode in the hands of the consumer and spread the flaming contents over the body of the user." This statement is totally inaccurate and false. It contradicts the tests performed by "P" and also documents the fact that Mr. "W" is limited with his lack of background in the areas of chemistry, chemical engineering, fires and explosions, deflagrations, flashbacks and why and how the container failed in the instant case. In addition, although Mr. "W" never addressed the subject, it is more likely than not, that no polymeric, glass or metal container could have withstood the deflagration that occurred.

What "W" is attempting to do is validate the defective scientific methodology used by the other plaintiff's experts that did not, in anyway, attempt to reproduce "A"'s description of his accident using an exemplar fireplace and the Diamond thixotropic product. The same group of "experts" chosen by the plaintiff are comparing the same inaccurate models and opinions with one another-but not versus the real world.

Furthermore, Mr. "W" never addressed the subject in which "A" irresponsible actions resulted in him squirting the contents of the Diamond Natural Fire Starter Gel in many areas on the carpet, the hearth, the walls and the firewood while it was burning. In addition he held the container for an extended period of time at a horizontal angle just like "P"'s test demonstrated before the vapor, emanating from the open container, caught on fire resulting in the deflagration of the container. ***This did not happen.***

Mr. "W" deficient background relating to the issues in this case is well beyond his capability of explaining what occurred in this case posing the subject incident and resulting injuries to "A". ***His reliance on the other experts inaccurate and inept attempt to reproduce the questionable testimony of "A" and his family rather than independently prove or disprove his case, defines his lack of***

experience and background in the technical issues associated in the subject case.

Fire and burning tests
Based upon over 50 years of experience in testing the flammability of a variety of solid and liquid chemical formulations in addition to testing the subject formulation in a simulated wood fireplace, it is the opinion of the undersigned, with a reasonable degree of safety engineering and scientific certainty, that "A" did not testify in his deposition as to his actual behavior prior to and up to the time he was injured. There is no question, based upon the testing by the undersigned and the plaintiff's expert's testing, that "A" did more than just squirt a little of the thixotropic alcohol into a burning fire. It is more likely than not that he squirted an excessive amount of the contents not only on the carpeting and the areas surrounding the fireplace such as the hearth, the back wall of the fireplace, the riser to the hearth, the walls, etc., but the straw that broke the camel's back was the excessive amount of formulation squirted on himself and the fire itself for an extended period of time until the vapors emanating from his body ignited due to the fire and heat radiating from the fireplace ending up igniting the vapors.

Experiments by the plaintiff's experts document the fact that an excessively large amount of squirts. The only way the vapors inside the container could be ignited was if "A" constantly squirted the contents out and air continued to enter the head space until there was sufficient air in the container to lower the concentration of the ethanol vapor to within the explosive limit so it could ignite. That was also verified in the video tape submitted by the plaintiff's expert.

In addition, the fact that "A" did not receive burns on the top of his head and back documents the fact that he did not raise the container above his head while the gel was coming out the bottom. That had to occur elsewhere such as on the or around the couch. (The resistance of the container from igniting and burning was well demonstrated in the fire test performed by the undersigned.)

Furthermore, this also verifies the fact that, in the opinion of the undersigned, "A"'s mother, "K", contrary to her sworn testimony under oath, never read the instructions to any of her children and never demonstrated that only a tablespoon amount of the contents should be placed on an unlit piece of wood in a wood stove. If she had, this would also document that "A" did not understand anything she attempted to warn "A" about and, as a 7 year old, he was irresponsible and disobedient.

. ***In addition to the above,*** **the undersigned squirted excessive amounts of the Diamond product into the burning fireplace from the front and from the top exposing the spout to well over 200 degrees Fahrenheit in the front and over 300 degrees Fahrenheit one foot from the flame from the top. All of these tests were videotaped and photographed. In both instances, the undersigned could not touch the closure on**

the top to close it. It was much too hot. In all cases, there was no indication of any flame front entering the container because it would be impossible for that to occur. Therefore, those tests disprove "A"'s account and testimony of what occurred prior to the subject incident.

The undersigned has consulted on several hundred burn cases to both children and adults over the last 44 years. Based on that experience and studies, I am able to convey that experience and observations to the following:

a. Explosions of natural gas, propane, gasoline and other flammable liquids, including ethanol, isopropanol, cause intense heat for a very brief time. Clothing, unless it ignites, is protective in flash burns. Hence, flash burns generally have a distribution involving all exposed skin, with the deepest areas facing the source of ignition. For the most part, flash burns reach progressive layers of the dermis in proportion to the amount and kind fuel that explodes.

b. In contrast to flash injuries, flame burns are invariably deep dermal if not full-thickness because of more prolonged exposure to intense heat.

c. Adding and squirting a voluminous amount of Diamond gel into the fireplace in all directions did not result in any flashback whatsoever. The heat emanating from the fireplace was too high in temperature for any human to withstand. If that is true, where was "A" actually positioned at the time his body caught on fire because of the large amount of gel all over his body? How did he really react once the gel all over his body caught on fire?

Based upon the extensive experiments performed by the undersigned on an exemplar fireplace squirting extensive amounts of gel on a burning fire, there was never a flashback experienced under any conditions. Any testimony regarding flashbacks squirting the Diamond product into a fire cannot be reproduced and would be incorrect.

d. With reference to the subject incident and "A"'s burns, it is highly probable based upon the amount of contents that he squirted into the fire, that he could have easily received a significant amount of his burn injuries from the gel that "A" squirted all over his body prior to getting on his knees in front of the fire.

. **Marshmallows**
There is some evidence of a white residue on a large metal fork in one of the photographs. *If "A" did roast one or more marshmallows and placed the fork down, this would then lend itself to "A" finishing the marshmallow(s) and then continuing to play with the Diamond product.* Squirting of the gel onto the carpet, hearth and other areas around the room, including his own body prior to kneeling before the unscreened fireplace and fire contributed to all of "A"'s burn injuries. All of the vapors emanating from his body was exposed to the heat from the fireplace. The temperature was more than enough to ignite the vapors. In fact, the temperature was more than enough to ignite most other solvents. No one ever found the bag of marshmallows "A" claimed he took downstairs to the fireplace.

Reconstruction of Subject Accident (Flammability Testing)
The undersigned conducted extensive fire tests in Boynton Beach, Florida by creating a exemplar fireplace using hickory wood chunks, paper and the Diamond Fire Starter Gel to start the fire. The following was observed:

a. **As the fire developed, the undersigned squirted large volumes of the gel onto the fire. No flash fire was observed. What was observed was a fire of higher intensity.**

b. **Large volumes of the Diamond product were squirted into the fire directly from the front. No flash fire was observed. What was observed was a fire of higher intensity.**

c. **A metal surface was covered with the gel and placed near the front of the fireplace. This was done to stimulate the gel covering "A"'s body that he previously squirted on himself. The surface burned with high intensity within a short period of time.**

d. **The surface of the 2" X 6" piece of wood, approximately 16" long was coated with the gel and placed along the floor surface adjacent to the front of the fireplace. Within a short period of time the vapors from the surface ignited and the gel on the surface burned.**

e. **Large volumes of the gel were squirted onto a blazing intense fire in the fireplace from a position directly above the fire and in front of the fire. No flashbacks occurred at any time. *The spout was exposed to temperatures ranging from 200 to 300 degrees Fahrenheit. The spout, containing the gel never caught on fire.* This contradicts all statements made by the plaintiff's experts and "A" in his sworn testimony.**

f. **The Diamond Fire Starter Gel container was placed directly in front of the fireplace. Nothing initially happened to the container. However, additional tests on the integrity of the container proved that it held up exceptionally well when**

exposed to temperatures over 200 degrees Fahrenheit as documented in the videotapes and photographs taken by the undersigned during the flammability and integrity testing of the container. ***The container was half filled and the only observation made while the container was exposed to over 200 degrees Fahrenheit was that the body of the container facing the fire expanded as shown in the videotapes and photographs.***

g. **The Diamond Fire Starter Gel container with approximately 40 cc's of gel then thrown into the fireplace. There was no explosion and it took an extensive amount of time before the container started to burn documenting the fact that the integrity of the design of the container was more than satisfactory.**

h. **The undersigned squirted large quantities of the Diamond Natural Fire Starter Gel from the opening in the top of the fireplace shown in the photographs and videotapes. The temperature opening of the top of the fireplace was well over** ***300*** **degrees Fahrenheit. It was so hot that the undersigned could not touch the top of the closure and seal the top with my bear hands. The temperature was well over ignition temperature of the ethanol vapor. In spite of that, no internal ignition was observed and no explosion occurred contrary to the testing and comments by Dr. "G" and "P". The latter two plaintiff experts did not, in any way, experimentally demonstrate their incorrect opinions.**

Testing the Integrity of the Diamond Natural Fire Starter Gel Plastic Container
The undersigned tested the integrity of a half filled container of the Diamond Natural Fire Starter Gel. The fire was started using the Diamond Natural Fire Starter Gel as the accelerant. Instead of a tablespoon of the material, approximately 1/3 of the container was sprayed on the wood nuggets and the vapors were ignited with a Scripto flame lighter. A continuous videotape was taken and the experiment was also documented with a series of photographs. A half filled container of the Diamond Natural Fire Starter Gel was placed approximately 14 inches away from the fire. The temperature of the heat emanating from the fireplace, at the point where the container was placed, was approximately 200 degrees Fahrenheit. No human being could withstand the temperature.

A half filled container of the Diamond Natural Fire Starter Gel was placed in front of the fireplace for over 30 minutes. After approximately 20 minutes, the area below the headspace that was facing the heat emanating from the fireplace swelled after approximately 25 minutes. There was no failure whatsoever and the bottom seam maintained its integrity contrary to the claims made by the plaintiff's experts. It is the opinion of the undersigned, with a reasonable degree of safety engineering and scientific certainty, that the plastic container is more than suitable for the formulation for the Diamond Natural Fire Starter Gel. In addition, unlike many formulations containing alcohol does not deform while in storage.

An additional important note was that the temperature emanating through the top hole in the cover of the garbage can where the undersigned squirted streams of the gel

into the fire below, was well over *300* degrees Fahrenheit. It was so hot, that the undersigned could not handle the closure for the top of the bottle after squirting the gel through the hole. Contrary to the claims made by the plaintiff's experts, there was no flame front that went through the opening in the spout that entered the internal headspace of the container. The reason was that the concentration of the ethanol vapor was above the upper explosion limit *and that the headspace was at the bottom of the container and not the top. When the container was placed in the upright position, and the spout left open until it cooled down, there was no ignition of the ethanol vapor escaping from the opening in the spout.*

The headspace above the half filled container contained mostly ethanol vapor. When exposed to the excessive heat in front of the exemplar fireplace, the gas expanded and the container responded by expanding as well. Contrary to the *defendants* allegations regarding the seam on the container, it would be predicted to anyone with knowledge of plastic containers, liquids and vapors within the container, that the most vulnerable area of the container when exposed to elevated temperatures would be the headspace within the container and not the bottom seam. *Even though this was foreseeable, it was also proven during the experiment by the undersigned.*

Furthermore, based on the demonstration by the undersigned, the plaintiff's claim that a leaked developed at the bottom seam of the container when the container was exposed in some unexplained way to their fireplace, is an impossibility. If it was exposed to excessive heat from the fireplace, the container would demonstrate an area of expansion above the headspace. The evidence in possession of the plaintiff documented that there was no expansion whatsoever. Therefore, it would be impossible for the accident to occur as explained under oath by "A", members of his family and the reports and experiments released by the plaintiff's experts.

- **The retail package provided sufficient barrier to prevent migration of the product or volatiles from the product.**

- **The Material Safety Data Sheet relating to this product was prepared in a reasonable manner and within all applicable standards.**

- **The delivery of this product, as delivered by PSC to Jarden/Hearthmark, was new and merchantable, free from any defects in workmanship or material, and conformed to applicable specifications, state of technology and the standard of care in the industry. Specifically, it is my opinion to a reasonable degree of engineering and scientific certainty, that the product met the specifications set forth by Jarden Home Brands / Hearthmark in the Performance Sepecifications for the Diamond Fire Starter Gel.**

Tests were conducted by the undersigned to determine the force required to open the closure that snapped in place. All measurements were over 8-pounds of force. That

would make the closure child resistant and very difficult, if not impossible for young children to open it up.

•

Conclusions

It is the opinion of the undersigned, with a reasonable degree of safety engineering and scientific certainty, that contrary to the allegations set forth by the plaintiffs and the plaintiffs expert's, that the cause of all of "A"'s injuries were directly related to his mother's *willful and intentional* disregard of the safety and welfare of her own son and had nothing to do with the formulation of the Diamond Natural Fire Gel or the integrity of the container. *The fact, at this time she is attempting to expiate her guilt by blaming her negligence on anyone of the defendants is a sham.* The conflicting statements by each member of her family, the EMTs and the medical specialist in the hospital more than document this opinion.

The flammable testing performed by the undersigned refuted and contradicted all of the allegations made by the plaintiff and by all of the plaintiff's experts. All of the testing performed by the plaintiff's experts was not realistic and did not relate, in any way, to the real world and what actually occurred at the time "A" was burned.

Summarily, the undersigned further opines that the Diamond Natural Fire Starter gel product is reasonably safe for its intended uses and reasonably foreseeable misuses, that the actions of PSC, Inc. in communicating with Jarden/Hearthmark and blending/packaging product were within all applicable standards of care, and PSC, Inc. did not cause or contribute to "A"s incident.

The experiments performed and opinions made by the plaintiff's experts did not represent the actual behavior of "A" prior to and up to the time of the actual incident. All of the experts' opinions were *"validated" by the plaintiff's handpicked group of alleged group of "experts" that compared* based on the results of inaccurate testing and questionable testimony of "A" and his family, *but* not versus the real world and real science. Even so, the Plaintiff's experts still could not replicate the version of events provided by the family.

7. Opinion

a. With over forty years of experience in the area of flammable fabrics, explosions, fires, children's burn injuries, warnings and instructions I cannot find liability in Diamond's Natural Fire Starter Gel.

b. "A"'s parents have a non-delegable duty to safeguard the safety welfare of "A.". They blatantly ignored that responsibility and allowed him to play unsupervised with a dangerous instrumentality. They are responsible for the consequences of their irresponsible behavior.

c. The undersigned has compassion for "A" and his injuries. However, the liability and the cause of his burn injuries are directly related to the disregard of "A"'s's safety and welfare by his parents.

C. J. Abraham,, PhD, JD, PE, CPC, FRSC, DEE, FTI, BCFE, DCEE, CChE, CChem has a varied background that can be found at: www.scientificadvisory.com. He also has over 45 patents in a variety of arenas as well. Many of the inventions were commercialized. The latest invention can be found on www.forcefieldheadbands.com

AUTHOR INDEX